Andreas Schlaich

Time-dependent spectrum analysis of high power gyrotrons

Karlsruher Forschungsberichte aus dem
Institut für Hochleistungsimpuls- und Mikrowellentechnik

Herausgeber: Prof. Dr.-Ing. John Jelonnek

Band 6

Time-dependent spectrum analysis of high power gyrotrons

von
Andreas Schlaich

KIT Scientific Publishing

Dissertation, Karlsruher Institut für Technologie (KIT)
Fakultät für Elektrotechnik und Informationstechnik, 2014
Referenten: Prof. Dr. rer. nat. Dr. h.c. Manfred Thumm
Prof. Dr.-Ing. Dr. h.c. Klaus Schünemann

Impressum

KIT Scientific Publishing

Karlsruher Institut für Technologie (KIT)
KIT Scientific Publishing
Straße am Forum 2
D-76131 Karlsruhe

KIT Scientific Publishing is a registered trademark of Karlsruhe Institute of Technology. Reprint using the book cover is not allowed.

www.ksp.kit.edu

Print on Demand 2015

ISSN 2192-2764
ISBN 978-3-7315-0375-0
DOI: 10.5445/KSP/1000046919

Foreword of the Editor

Gyrotrons are high power millimeter-wave oscillators, capable of delivering coherent continuous-wave (CW) power in the megawatt range. This makes them a key component of thermonuclear fusion experiments using magnetically confined plasmas, where they enable electron-cyclotron resonance heating (ECRH) and electron-cyclotron current drive (ECCD) for plasma stabilization. For example, the ECRH system of the stellarator experiment Wendelstein 7-X requires 10 MW of heating power at 140 GHz; the ITER facility is currently projected with a 24 MW ECRH system at 170 GHz.

As for every oscillator, the most significant information on the quality of operation of every gyrotron can be deduced from the examination of RF output spectrum. Typical properties of interest include the operating frequency, the existence of parasitic oscillations and their behavior, mode switching, and modulation of the operating frequency. Since many of these properties can appear in combination or interdependence, information about the stability and evolution of these effects during the pulse is highly desired.

Dr.-Ing. Andreas Schlaich presents a novel measurement system, which permits time dependent spectral measurements within a large bandwidth, dynamic range and unambiguous RF indication in the entire D-Band (110–170 GHz). To date the system is unique within the gyrotron community, and facilitates the investigation of transient spectral effects with unprecedented detail. The investigation of parasitic RF oscillations in high-power gyrotrons, combining theory, simulation and experiment, forms another pivotal point of the presented work. Here a comprehensive classification of the various possible parasitic signals is established, including paths for systematic diagnosis. The results form an important contribution to the ongoing discussion about after-cavity interactions (ACI) in gyrotrons.

In addition, the measurement system was used to investigate mechanisms inside the gyrotron which lead to inherent frequency tuning at the start of long RF pulses. By combining precise frequency measurements with an analytically derived model, more information on the ionization-based space charge neutralization effects and the thermal expansion of the gyrotron resonator was obtained with a numerical fitting method.

Dr.-Ing. Andreas Schlaich has provided the gyrotron community with a new tool for experimental gyrotron monitoring and diagnosis, which allows for more rigorous verification of simulation results. The overall approach of decreasing the distance between the domains of experiment and simulation is well perceived throughout this work. We are sure that it will be met with large resonance.

Time-dependent Spectrum Analysis of High Power Gyrotrons

Zur Erlangung des akademischen Grades eines

DOKTOR-INGENIEURS

von der Fakultät für
Elektrotechnik und Informationstechnik
des Karlsruher Institut für Technologie (KIT)

genehmigte

DISSERTATION

von

Dipl.-Ing. Andreas Schlaich

geb. in Freudenstadt

Tag der mündlichen Prüfung:	12.12.2014
Hauptreferent:	Prof. Dr. rer. nat. Dr. h.c. Manfred Thumm
Korreferent:	Prof. Dr.-Ing. Dr. h.c. Klaus Schünemann

Kurzfassung

Die vorliegende Arbeit behandelt die Untersuchung dauerbetriebsfähiger Hochleistungsoszillatoren im Millimeterwellenbereich (sog. Gyrotrons) durch Spektralmessungen.

Im Zentrum steht die Entwicklung eines breitbandigen Messsystems zur zeitvarianten Spektralanalyse mit einem hohem Dynamikbereich von 50–60 dB, welches den Frequenzbereich 100–170 GHz mit einer Echtzeitbandbreite von 6–12 GHz abdeckt. Trotz Heterodynempfangs mit harmonischen Mischern erzeugt das *Pulse Spectrum Analysis* (PSA) System unter Anwendung einer speziellen Rekonstruktionsmethode frequenzeindeutige HF-Spektrogramme. Mit Hilfe dieser Messmittel wurde eine Vielzahl an spektralen Effekten in Hochleistungsgyrotrons untersucht, wie z.B. Hauptmodenwechsel, Frequenzverstimmung im Langpulsbetrieb, Niederfrequenzmodulation, und das Verhalten während eines Vakuumfensterüberschlags.

Im Rahmen einer Untersuchung zu parasitären RF-Oszillationen in Gyrotrons für den Stellarator Wendelstein 7-X (W7-X) konnte gezeigt werden, dass sogenannte *after-cavity oscillations* in modernen Hochleistungsgyrotrons auftreten können und vermutlich die Ursache der bei den W7-X Gyrotrons beobachteten unerwünschten Schwingungen darstellen. In Kombination mit systematischen Parameterstudien im Experiment kamen hier die Dispersionsanalyse im Brillouindiagramm sowie umfangreiche Gyrotron-Wechselwirkungssimulationen zur Anwendung.

Das PSA Messsystem wurde außerdem eingesetzt, um die inhärente Frequenzverstimmung durch Wärmeausdehnung und elektrostatische Neutralisierungsvorgänge im Langpulsbetrieb zu untersuchen. Durch makroskopische Modellierung der gasdynamischen und ionisierenden Prozesse konnten in Kombination mit einer Fitting-Methode deren Zeitabhängigkeiten untersucht werden. Hierbei ergaben sich unter

Anderem Hinweise darauf, dass in W7-X Gyrotrons nur etwa 60% der elektrostatischen Spannungsabsenkung neutralisiert werden, an Stelle der üblich angenommenen 100% im stationären Zustand.
Die systematische messtechnische Untersuchung der genannten Effekte wird durch umfassende Untersuchungen in Simulation und Theorie ergänzt. Besonderer Wert wurde auf die Schaffung von Verknüpfungspunkten zwischen Experiment und Simulation gelegt; unter Anderem wurde eine ausführliche und systematische Parameterstudie mit mehreren führenden europäischen Gyrotron-Wechelwirkungs-Simulationswerkzeugen im Kontext konkreter Messergebnisse durchgeführt.

Abstract

In this work, an investigation of vacuum electronic oscillators capable of generating multi-megawatt continuous wave output power in the millimeter-wave range (so-called gyrotrons) through spectral measurements is presented.

The centerpiece is the development of a measurement system with a high dynamic range (50–60 dB) for time-dependent spectrum analysis, covering the frequency range 100–170 GHz with instantaneous bandwidths of 6–12 GHz. Despite relying on heterodyne reception through harmonic mixers, the *Pulse Spectrum Analysis* (PSA) system maintains RF unambiguity in the spectrogram output through the application of a novel RF reconstruction technique.

Using the new possibilities, a wide range of spectral phenomena in gyrotrons has been investigated, such as cavity mode jumps, low-frequency modulation, frequency tuning in long pulses and the spectral behavior during the presence of an RF window arc.

A dedicated investigation on parasitic RF oscillations in W7-X gyrotrons combining several analysis techniques led to the conclusion that after-cavity oscillations can be physical reality in high power gyrotrons, and are the probable cause for the undesired signals observed. Apart from systematic parameter sweeps using the PSA system, an analytical dispersion analysis in the Brillouin diagram was applied, and numerical gyrotron interaction simulations of unprecedented extent were conducted.

Furthermore, the improved frequency measurement capabilities were employed to analyze the frequency tuning through thermal expansion and electrostatic neutralization caused by ionization inside the tube in long-pulse operation. By macroscopically modeling the gas dynamics and ionization processes in combination with a fitting process, the time dependences of the two processes could be investigated. In doing

so, indication was found that the neutralization in W7-X gyrotrons amounts to only 60% of the electrostatic depression voltage, instead of 100% as widely believed for stationarity.
The systematic experimental characterization of these effects was complemented by thorough investigations using theoretical means and simulations. Here a focus was put on bridging the often too-wide gap between the worlds of simulation and experiment, including the derivation of connections between the quantities involved and rigorous benchmarking. For this, also an extensive and systematic comparison study of multiple major European gyrotron interaction simulation codes was performed and compared with measured values.

Contents

Symbols, variables and abbreviations **xi**

1 Introduction **1**

1.1 Gyrotrons and their applications 1

1.1.1 Gyrotrons as millimeter-wave oscillators 1

1.1.2 Fusion science application 2

1.1.3 Further gyrotron applications 4

1.2 Spectral measurement challenges in gyrotrons 4

1.2.1 Determination of the operating cavity mode . . 5

1.2.2 Frequency tuning during long pulses 6

1.2.3 Linewidth and short-time stability 6

1.2.4 Low-frequency oscillations (LFOs) 6

1.2.5 Parasitic RF oscillations (POs) 7

1.3 Review of spectrum analysis techniques 8

1.3.1 Frequency conversion 8

1.3.2 Measurement and receiver types 9

1.4 Scope of this work 12

1.4.1 Measurement system 12

1.4.2 Investigations 13

2 Gyrotron setup and behavior **15**

2.1 Electron-optical system 15

2.1.1 Electron beam and macroscopic properties . . . 15

2.1.2 Adiabatic quantities 18

2.1.3 Electrostatic voltage depression effect 19

2.2 Microwave fields and gyrotron interaction 20

2.2.1 Waveguide modes 20

2.2.2 Interaction 21

2.2.3 Basic beam-wave synchronization 22

2.3 RF-relevant geometry . . . 26
2.3.1 Gyrotron cavity . . . 26
2.3.2 Cathode to cavity . . . 26
2.3.3 Cavity to output window . . . 27
2.4 Behavioral aspects . . . 28
2.4.1 Mode competition and starting currents . . . 28
2.4.2 Schottky-effect . . . 30
2.4.3 Pulse characterization from duration . . . 30
2.4.4 Long-line effect . . . 33

3 Measurement and simulation **37**
3.1 Gyrotron experiments . . . 37
3.1.1 Experimental settings . . . 37
3.1.2 Frequency and spectrum measurement . . . 38
3.1.3 Power measurement . . . 40
3.1.4 Example measurement . . . 44
3.2 Simulation techniques . . . 45
3.2.1 Beam parameters . . . 45
3.2.2 Gyrotron interaction analysis . . . 47
3.3 Benchmarking of experiment and simulation . . . 52
3.3.1 Frequency calculation results . . . 53
3.3.2 Self-consistent simulation codes and variants . . 54
3.3.3 Scenario . . . 56
3.3.4 Sample start-up with SELFT . . . 59
3.3.5 Sample start-up with EVRIDIKI . . . 61
3.3.6 Benchmark results . . . 62
3.3.7 Discussion . . . 64
3.3.8 Benchmark summary . . . 67

4 Pulse spectrum analysis **69**
4.1 General hardware setup . . . 69
4.2 Considerations on dynamic range and bandwidth . . . 71
4.2.1 IF receiver . . . 71
4.2.2 Mixers and local oscillators . . . 73
4.2.3 Total dynamic range . . . 75

4.3 IF data processing 76
4.4 RF spectrum reconstruction 77
4.4.1 Principle 77
4.4.2 Implementation 79
4.4.3 Effective dynamic range 82
4.5 Reconstruction robustness criteria 82
4.5.1 Coincidental false-positive validations 83
4.5.2 Deterministic false-positive validations 83
4.5.3 Estimation of reception possibilities 84
4.6 Summary on parameter choices 89
4.6.1 Desired mixing harmonics 89
4.6.2 LO frequency shift Δf_{d} 90
4.6.3 FFT / STFT configuration 91
4.7 Spectrogram post-processing and time extension 91
4.7.1 Frequency extraction 91
4.7.2 Operation with undersampling 93

5 Primary measurement results **97**
5.1 Nominal cavity mode appearance 97
5.1.1 Typical short-pulse operation 98
5.1.2 Start of long pulse operation 99
5.1.3 Long-pulse operation monitoring 101
5.2 Cavity mode instabilities 104
5.2.1 Power supply voltage ripple 104
5.2.2 Short-time stability 105
5.3 Cavity mode competition and switching 107
5.3.1 Cavity mode switch in a W7-X gyrotron 107
5.3.2 Parallel cavity modes and LF modulation in a W7-X gyrotron 108
5.3.3 Parallel cavity modes and LF modulation in the 2 MW coaxial gyrotron 110
5.4 Spectral distortions during a window arc 112
5.4.1 Observed RF behavior 112
5.4.2 Discussion of RF behavior during window arc . 114
5.4.3 Causal investigations 115

6 Parasitic oscillations **117**
6.1 Background and classification 117
6.1.1 General deleterious effects 117
6.1.2 Before-cavity interactions (BCIs) 119
6.1.3 After-cavity interactions (ACIs) 121
6.1.4 Diagnosis of BCIs and ACIs 125
6.2 Measurements 126
6.2.1 Intra-pulse behavior and spectral appearance . 127
6.2.2 Sweep over cathode voltage 132
6.2.3 Sweep of magnetic flux density 136
6.2.4 Scatter analysis 141
6.3 Brillouin-Analysis 143
6.3.1 Fundamentals 143
6.3.2 Beam parameters along the z axis 144
6.3.3 Applicability and scope 146
6.3.4 Method summary 147
6.3.5 Investigation of the W7-X gyrotron parasites . . 148
6.3.6 Result summary 155
6.4 Comparative simulations 157
6.4.1 Simulation setup 157
6.4.2 Sweep over cathode voltage 164
6.4.3 Sweep over beam current 168
6.5 Summary 170

7 Electrostatic and thermal effects during long pulses **173**
7.1 Overview 173
7.1.1 Relevance to gyrotron operation 173
7.1.2 Investigation approaches 174
7.1.3 Connection of physical processes 176
7.2 Electrostatic neutralization investigations 179
7.2.1 Basic ionization process 179
7.2.2 Neutral gas in the gyrotron 180
7.2.3 Ionization cross sections 183
7.2.4 Ion confinement and energies 185
7.2.5 Particle densities and flows 190

7.2.6 Time development scenarios 192
7.3 Time dependence fitting 195
7.3.1 Operating point transition 195
7.3.2 Simulated parameter map 196
7.3.3 First-order approach without outgassing 198
7.3.4 Outgassing revisited 205
7.3.5 Sample results 208
7.4 Summary 210

8 Conclusions **213**
8.1 Summary 213
8.2 Outlook 215

A Digital signal analysis by STFT **219**
A.1 Fourier transform of discretized signals 219
A.1.1 Aliasing 221
A.1.2 Power spectra 222
A.1.3 Window functions 222
A.2 Dynamic range 223
A.2.1 Time domain 223
A.2.2 Frequency domain 225
A.3 Time-frequency analysis 226
A.3.1 General considerations 226
A.3.2 Discrete short-time Fourier transform 227
A.3.3 STFT time step / overlap factor 227
A.3.4 Interpretation of spectrograms 228

B PSA system technical supplement **231**
B.1 Control software 231
B.2 Data format 232
B.3 Components and detailed schematic 235

C Gas kinetics and pressure **237**
C.1 Ideal gas conditions and molecular flow rules 237
C.1.1 Basic relationships 237
C.1.2 Molecular flow 239

C.2 Description of molecular gas flows 239
C.3 Connections between gas vessels 240
C.4 Gas particles and vessel walls 241
C.5 Vacuum pump basics 243
C.5.1 Pump types 243
C.5.2 Characterization 243
C.6 Ion getter pumps (IGP) 244
C.6.1 General information 244
C.6.2 Pressure measurement 244
C.6.3 W7-X gyrotron pumps 245

D Extended simulation results **247**
D.1 SELFT 248
D.2 EVRIDIKI 249
D.3 EVR-PIC 250

E Afterword **251**

Bibliography **255**

List of own publications **279**

Index **285**

Used symbols, variables, units and abbreviations

Symbols, arabic variables and units

A	geometrical area or cross-section
$\vec{B}$	magnetic flux density vector
$B_\mathrm{r}, B_\phi, B_\mathrm{z}$	components of the magnetic flux density vector in cylindrical coordinates
b	vertical A/D converter resolution, expressed as the number of bits
C	gas flow conductivity, [volume/time]
c_0	speed of light in vacuum
D	dynamic range
d	diameter
dB	decibel (base 10 logarithmic pseudo-unit)
dBc	decibel, relative to carrier level (resp. dominant signal)
dBm	decibel, normalized to the reference power of $1\,\mathrm{mW}$
Δf	frequency difference or bandwidth, i.e. $\Delta f = f_2 - f_1$
Δf_IF	IF bandwidth
Δf_d	desired frequency difference for RF reconstruction
Δt	time difference, i.e. $\Delta t = t_2 - t_1$
δf	frequency resolution
δt	time resolution
$\vec{E}$	electric field strength vector
$E_\mathrm{r}, E_\phi, E_\mathrm{z}$	components of the electric field strength vector in cylindrical coordinates
$\mathrm{E}[X]$	expected value of random variable X (statistics)
e	elementary charge
$\mathcal{F}\{x\}$	Fourier transform of x

f	frequency *or* probability density function
f_{cut}	cut-off frequency
f_{cyc}	relativistic electron cyclotron frequency
f_{N}	Nyquist frequency
$Ge_{m,n}$	beam-wave coupling coefficient of the mode $\mathrm{TE}_{m,n}$
g_i	ionization rate
$H(x)$	Heaviside-function
h	mixing harmonic, $h \in \mathbb{N}$
I	electronic current
k	integer number *or* wave propagation constant
L	load factor (for A/D converter)
l	length
m_{e}	electron rest mass
N, n	general integer numbers (e.g. of samples)
N_{x}	number of particles in volume (x: b̲eam, g̲as, i̲ons)
n_{x}	particle density in volume (x: b̲eam, g̲as, i̲ons)
P	power
p	pressure *or* momentum
q	statistical quantile *or* gas flow *or* charge
q_{N}	particle flow, $[1/\mathrm{time}]$
q_{pV}	pV-flow or throughput, $[\mathrm{pressure} \cdot \mathrm{volume}/\mathrm{time}]$
q_{V}	volume flow, $[\mathrm{volume}/\mathrm{time}]$
$R(f,t)$	RF spectrogram (reconstructed)
r	radial coordinate *or* geometrical radius *or* reflection factor
r_{b}	electron beam guiding center radius
r_{c}	radius of cylindrical cavity section
r_{w}	wall radius
S	pumping speed (defined as a volume flow)
$S(f,t)$	IF spectrogram
SNR	signal-to-noise ratio
s	power correction factor
$s(t)$	time-harmonic signal
T	temperature
t	time

V	electric voltage *or* volume
v	velocity in m/s
$w(t)$	scaling function
x_{OV}	overlap factor for STFT
Z	impedance *or* integer Nyquist zone index
z	axis or coordinate in cylindrical system

Greek variables

α	pitch factor, i.e. the velocity ratio $\alpha = \beta_{\perp}/\beta_{\|\|}$
β	normalized velocity, $\beta = v/c_0$
$\chi_{m,n}$	eigenvalue of the cylindrical waveguide mode $\mathrm{TE}_{m,n}$
ε_0	electric permittivity of vacuum
ϵ	residual error (RMS)
γ	Lorentz factor
λ	wavelength *or* change rate of time-dependent process
Ω_0	non-relativistic angular electron cyclotron frequency
Ω_{r}	relativistic angular electron cyclotron frequency
ω	angular oscillation frequency, $\omega = 2\pi f$
σ	standard deviation *or* ionization cross section
τ	alternative time variable *or* process time constant
ζ	voltage depression neutralization fraction

Abbreviations

A/D	analog-to-digital
A.U.	arbitrary units (in axis labels)
ACI	after-cavity interaction
AFDR	aliasing-free dynamic range
BCI	before-cavity oscillation
CDR	coupled dispersion relation
CPW	cavity potential well
CW	continuous-wave
DFT	discrete Fourier transform

ECCD	electron-cyclotron current drive
ECR(H)	electron-cyclotron resonance (heating)
ENOB	effective number of bits
FAT	factory acceptance test
GS	gigasamples (10^9 samples)
HSI	high-side injection for frequency mixing, i.e. $f_{\mathrm{LO}} > f_{\mathrm{RF}}$
HVPS	high voltage power supply
IF	intermediate frequency, frequency at receiver unit of a heterodyne system
IGP	ion getter pump, a type of vacuum pump
KIT	Karlsruhe Institute of Technology
LFM	low-frequency modulation
LFO	low-frequency oscillations
LO	local oscillator
LP	long-pulse (operation)
LSI	low-side injection for frequency mixing, i.e. $f_{\mathrm{LO}} < f_{\mathrm{RF}}$
MS	megasamples (10^6 samples)
PIC	particle-in-cell
PSA	Pulse Spectrum Analysis (system)
RMS	root-mean-square
RF	radio-frequency (at measurement system front-end)
SFDR	spurious-free dynamic range
SI	International System of Units (*Le Système international d'unités*)
SNR	signal-to-noise ratio
SP	short-pulse (operation)
SR	sampling rate (samples per second)
STFT	short-time Fourier transform
TE	transverse electric
TM	transverse magnetic
UDR	uncoupled dispersion relation
W7-X	Wendelstein 7-X (stellarator plasma experiment in Greifswald, Germany)

1 Introduction

In this Chapter an overview of the gyrotron as a high-power millimeter-wave oscillator and its applications is given. After revisiting a selection of typical diagnostic challenges in fusion gyrotron testing and development, the available measurement techniques of the field are reviewed.
Based on this, the scope of this work is described in the last Section.

1.1 Gyrotrons and their applications

1.1.1 Gyrotrons as millimeter-wave oscillators

The generation of high continuous wave (CW) power at microwave and millimeter-wave frequencies is generally limited by breakdown field strengths and allowable loss power density of the source. For solid-state devices, typical field strengths in the diffusion zone are the limit, but since e.g. in field-effect transistor technology the gate length scales with the wavelength, cooling problems introduce further difficulty for higher frequencies.
Vacuum electronic devices have fewer field effect problems and inherently better cooling possibilities, but are still limited in power density by their RF structure dimensions: Classical tubes such as klystrons, magnetrons or travelling-wave tubes (TWT) are so-called slow-wave devices, a term which refers to matching the phase velocity of the electromagnetic field to the velocity of the electron beam for a one-dimensional interaction along the beam axis. This approach requires slowing structures in the same size as the RF wavelength, e.g. fundamental mode resonators or the small-radius helix in TWTs.
On the contrary, gyro-devices use a strong magnetic field to define the resonance frequency and to induce helical trajectories in the electron

beam, causing the beam-wave interaction to take place in the transverse plane. This allows for very simple oversized resonators and removes the constraint of wavelength-size elements in such fast-wave devices.

The gyrotron (i.e. gyro-monotron or gyro-oscillator) is a powerful oscillator based on this concept. Its typical application range spans from a few tens of GHz up to the THz range, while in the millimeter-wave domain it allows the generation of multi-megawatt output CW power [NTP14]. A more detailed discussion on the operation of gyrotrons can be found in Chapter 2.

1.1.2 Fusion science application

In the last decades the necessity for energy sources alternative to fossil or nuclear fission based technologies has become increasingly clear. While the advent of so-called renewable energy sources like solar and wind power is both necessary and encouraging, their capability of shouldering tomorrow's worldwide demands for electrical energy remains under discussion [Mac11].

Among the further promising alternative energy sources for the next decades are power plants using thermonuclear fusion in magnetically confined plasmas. Like the sun, they draw their energy from the fusion of light atomic nuclei.

Already very early investigations of the possible fusion energy yield (e.g. using the Lawson-criterion, [Law57]) indicate that only fusion of the lightest nuclei, i.e. hydrogen and helium isotopes, is suitable for economic energy production; the overall most promising fusion process has been found to be the one between deuterium and tritium ("D-T-Fusion"). For such a reactor to operate, the energy required for overcoming the Coulomb repelling force between the positively charged particles must be available from the collective kinetic energy of the nuclei; it is in the order of $10\,\mathrm{keV}$ ($\approx 116\,\mathrm{Mio.\ K}$).

The most common setup enabling this reaction is a toroidal plasma vessel with magnetic confinement, where the reactants are heated by external power sources to reach the fusion conditions [She94].

One of the dominating techniques is electron-cyclotron resonance heating (ECRH), which is based on resonant absorption of RF energy by the plasma electrons around their cyclotron frequency

$$\Omega_{\mathrm{r}} = 2\pi f_{\mathrm{cyc}} = \frac{eB}{m_{\mathrm{e}}\gamma} \approx 2\pi \frac{28\,\mathrm{GHz}\cdot B/\mathrm{T}}{\gamma} \tag{1.1}$$

with e: elementary charge, $1.602 \cdot 10^{-19}\,\mathrm{C}$
m_{e}: electron rest mass, $9.109 \cdot 10^{-31}\,\mathrm{kg}$.

For typical fusion experiments these frequencies are found in the range 100–200 GHz [HAB+07, EBB+07, LAB+02] . The required heating powers of 10–24 MW gave rise to ECRH systems consisting of multiple gyrotrons with powers around 0.5–2 MW per unit.
The energy absorbed by the electrons is quickly being spread all over the plasma due to the high electron mobility, and is gradually transferred to the plasma ions through Coulomb collisions.
Apart from the primary heating purpose, ECR systems are also found to be suitable for other critical applications concerning the maintenance of plasma stability. Controlled ECR current drive in the plasma has been established many years ago, leading not only to the possibility of large scale non-inductive bulk current drive, but also to the prospect of local plasma manipulations [PFC+09, ZGG+01]. These are demanded for the control of complex destabilizing plasma phenomena, e.g. magnetic island formation, sawtooth instabilities and so-called neo-classical tearing modes, which can lead to problematic plasma disruptions. However, these applications demand more challenging features from the gyrotrons, such as multi-frequency operation, fast frequency tunability or power modulation.
The gyrotrons discussed within this work are developed by the Institute for Pulsed Power and Microwave Technology (IHM) of the Karlsruhe Institute of Technology (KIT) and are intended for the application in fusion experiments as described above, however there are other fields which also make use of the gyrotron technology.

1.1.3 Further gyrotron applications

The primary other application for CW gyrotrons is in microwave material processing at various frequencies, as low as 30 GHz [LFT+99] and as high as 300 GHz [AMS+11], for advanced sintering of ceramics or modern composite materials processing.
Another field is the so-called dynamic-nuclear polarization (DNP) enhanced nuclear magnetic resonance (NMR) spectroscopy, which, after its discovery [Ove53], highly relied on the advancement of the utilized microwave sources [GP10]. Today an external gyrotron signal, for example in the range 200–600 GHz [BHK+07], is used to increase the dynamic range of general NMR methods considerably, and to radically reduce measurement times.
Applications of gyrotrons in the radar and military field are not uncommon, as powerful millimeter-wave sources allow low tracking angles [LND+08], improved ballistic missile defense and space object detection [CMO00]. Since control over the signal shape is important, here typically gyro-amplifiers (e.g. gyro-klystrons) instead of gyro-oscillators are used. Other military projects include gyrotrons as part of non-lethal directed-energy weapon systems [Hac06], and as an alternative to laser-based systems.
More exotic are the applications for material melting [WSC+07], rock drilling [WEO13], the concept of wireless power transmission [Dic13] or NASA's millimeter-wave thermal launch system [MP12, Par06].

1.2 Spectral measurement challenges in gyrotrons

As for every oscillator, the most significant information on the quality of operation can be deduced from examination of the gyrotron's RF output spectrum. Typical properties of interest include the output frequency, existence of parasitic oscillations and their behavior, mode switching, low-frequency oscillations, and modulation of the output frequency. Since many of these properties can appear in combination or interdependence, information on the stability and evolution of these effects during the pulse is desired.

In the following, the relevant spectral measurement challenges and their impact on the requirements for diagnostic systems are described; a limited anticipation of relations discussed in Chapter 2, however, is necessary for some points to be made.[1]
The application focus is on fusion gyrotrons, but of course most concepts also apply for other gyrotron types. However, most of the investigations in the presented work were made in the context of the TH1507 series gyrotron development for the stellarator Wendelstein 7-X (W7-X). More information on the W7-X gyrotron development can be found in [DBC$^+$99], and a description of the W7-X ECRH system is given in [EBB$^+$07].

1.2.1 Determination of the operating cavity mode

Since the gyrotron resonator radius is known, the operating cavity mode can basically be identified by frequency measurement. The required accuracy depends on the type of gyrotron and the mode spectrum density.[2]
Typical candidates for identification against the nominal mode are azimuthal neighbors. In the W7-X gyrotron with its nominal cavity mode $TE_{28,8}$, this is typically $TE_{27,8}$. Measurement resolution of a few GHz is already enough to separate these modes.
Of course the mode spectrum is much denser, and there are cases which are harder to distinguish. The separation of co- and counter-rotating modes for example requires a resolution better than 100 MHz.
For simulation tool assessment, higher accuracies are needed. Here, at least 10 MHz frequency resolution are necessary to make meaningful comparisons (compare Section 3.3).

[1]The index section at the end of the document is useful for finding the corresponding definitions, if necessary.

[2]In gyrotrons with axial output, more direct access to the operating cavity mode is given than for gyrotrons with quasi-optical mode converters. In this case, so-called *k*-spectroscopy (see [KM88, BKK$^+$88]) can provide more detailed information about the mode mixture in the device. Axial output coupling, however, limits the achievable RF CW power far below 1 MW, leading to the establishment of transverse quasi-optical coupling as the high power state-of-the-art.

1.2.2 Frequency tuning during long pulses

During the first few hundred milliseconds of longer pulses, the operation frequency of the cavity mode typically tunes over a range of up to 500 MHz due to deterministic electrostatic potential shifts and the thermal expansion of the cavity (see e.g. [DBC+99, WGL+08]).
For the proper understanding of such effects, detailed knowledge of the time dependence is necessary. This calls for a frequency span higher than the expected tuning range and resolutions better than 10 MHz. Of course also a time-dependent measurement with the capability for acquisition times comparable with the duration of the tuning effect, i.e. up to 1 s, is required.

1.2.3 Linewidth and short-time stability

In fusion gyrotrons, the linewidth and noise properties are not of concern for the primary applications of ECRH and non-inductive current drive. However, for a gyrotron developer, the knowledge of these properties is desirable for benchmarking simulation tools, and investigating the potentially deleterious effects of things like extreme beam parameter values or misalignment in the magnetic system.
Furthermore, the fluctuation of external parameters such as the cathode, anode and collector voltages can cause significant short-time instability of the oscillations, which can be characterized over the line width. Here naturally high frequency resolutions in the MHz or even kHz range are useful, in combination with parameter flexibility.

1.2.4 Low-frequency oscillations (LFOs)

This area of investigation loosely covers a large number of effects, whose common property is that in comparison with the gyrotron interaction, the oscillation appears at comparably low frequencies, at most a few GHz. The radiated power from LFOs can be as much as multiple kW, creating personnel safety issues as well as malfunctions in neighboring equipment [LCC+00]. There are two basic ways of detecting LFOs, either by probing the exterior of the gyrotron with

radio pickup antennas, or by diagnosing modulation patterns on the RF-measured millimeter-wave signal.
Possible origins for LFOs are independent oscillations of trapped agglomerated charges inside the gyrotron [Tsi01], or parasitic non-gyro-type beam-wave interactions outside the cavity. If such independent oscillations are taking place in the area between the gun and the cavity, they can easily modulate the electron beam axially, causing a modulation imprint on all downstream oscillations.
Modulation of cavity modes, however, can have a variety of sources: For example also internal or external power reflections can cause coherent modulation of an operating mode through the load-pull effect [AJCN92]. Furthermore, there exists the phenomenon of self- or automodulation, which causes the amplitude to be modulated [Nus85, ADRS01].
Normal RF measurements will only detect LFOs in the form of modulation, for which an adequate bandwidth in the GHz range would be suitable. Without additional LF pickup equipment, one cannot easily determine the origin of occurring LF modulation.

1.2.5 Parasitic RF oscillations (POs)

Generally this term is used to describe undesired gyrotron-type interactions. These appear typically outside of the desired interaction structure, the cavity. However, additional oscillations inside the cavity are also often termed this way, see e.g. [ZZN79, GBD$^+$07]. In this work, we will primarily use this term for oscillations outside of the cavity, while describing unwanted in-cavity oscillations as competing interactions.
The primary effect of out-of-cavity parasitic oscillations lies in a distortion of the electron beam in the form of parameter spread increase. If it happens between the gun and the cavity, this deteriorates the performance of the cavity mode and leads to less predictive behavior of the gyrotron, for example by changing stability parameter ranges of the cavity modes. Behind the cavity, the distortion can still lower the achievable efficiency by limiting the applicable collector depression

voltage. A more elaborate discussion of parasitic RF oscillations is given in Chapter 6, including experimental and simulation studies in W7-X gyrotrons.
Parasitic oscillations have been documented during most high-power gyrotron development projects worldwide (e.g. [GDF+10, SSH+01, PAH+98]), and are a common problem in all high-power vacuum electronic devices. The goal of suppressing them has a major impact on design strategies, which can only be reliably achieved when the type and place of origin of the unwanted oscillation are known. Then, appropriate changes in the electron beam configuration or the RF geometry can be made.
Since initially the frequency and power level of POs is unknown, for detection, characterization and exclusion a high measurement bandwidth and dynamic range are desired. The behavior of a parasite in the context of cavity mode fluctuations, or vice versa, can be of interest; this requires the possibility of time-dependent and concurrent monitoring of different spectral areas, as the parasite can be easily 10 or 20 GHz away from the cavity mode. Parasites often behave less regularly than the cavity mode, and with limited reproducibility, so ambiguities in the single-shot frequency determination should be avoided.

1.3 Review of spectrum analysis techniques

In the following, the relevant methods for spectrum analysis in the millimeter-wave domain are described, focusing on their applicability to the challenges presented above.

1.3.1 Frequency conversion

All receiver types can be extended to a wide frequency range through heterodyne conversion. Here a high frequency input range (*radio frequency*, RF) is converted by a mixing element pumped with a *local oscillator* (LO) signal, into a lower (*intermediate frequency*, IF) range. Especially in the millimeter-wave domain, these mixers often operate on harmonics $h \cdot f^{\mathrm{LO}}$.

Depending on the relative position of f^{LO} and f^{RF}, the frequencies are converted following

$$f^{\mathrm{IF}} = \begin{cases} f^{\mathrm{RF}} - h \cdot f^{\mathrm{LO}} & \text{for } f^{\mathrm{RF}} > h \cdot f^{\mathrm{LO}} \leftrightarrow \text{low side injection (LSI)} \\ h \cdot f^{\mathrm{LO}} - f^{\mathrm{RF}} & \text{for } f^{\mathrm{RF}} < h \cdot f^{\mathrm{LO}} \leftrightarrow \text{high side injection (HSI)} \end{cases} .$$

Thus, all signals received through the available mixing harmonics and injection sides will be situated in the IF range. When attempting to determine the original RF position of frequency components found in the IF range, one faces substantial ambiguity, since neither the mixing injection side nor the acting harmonic are known without a-priori knowledge. For a number of N possible mixing harmonics, the ambiguity is $2N$-fold.
For example, an IF signal at $f^{\mathrm{IF}} = 2\,\mathrm{GHz}$ received with $f^{\mathrm{LO}} = 14\,\mathrm{GHz}$ can originate from an RF signal at any of the following positions:

- ...
- $f^{\mathrm{RF}}_{h=8,\ \mathrm{LSI}} = 8 \cdot f^{\mathrm{LO}} + f^{\mathrm{IF}} = 114\,\mathrm{GHz}$ or
- $f^{\mathrm{RF}}_{h=8,\ \mathrm{HSI}} = 8 \cdot f^{\mathrm{LO}} - f^{\mathrm{IF}} = 110\,\mathrm{GHz}$ or
- $f^{\mathrm{RF}}_{h=9,\ \mathrm{LSI}} = 9 \cdot f^{\mathrm{LO}} + f^{\mathrm{IF}} = 128\,\mathrm{GHz}$ or
- $f^{\mathrm{RF}}_{h=9,\ \mathrm{HSI}} = 9 \cdot f^{\mathrm{LO}} - f^{\mathrm{IF}} = 124\,\mathrm{GHz}$ or
- ...

1.3.2 Measurement and receiver types

Different measurement techniques are often distinguished by the respective signal domain of operation [TWK98]:

- time domain (amplitude vs. time)
- frequency domain (amplitude vs. frequency)
- modulation domain (frequency vs. time).

As these domains are interconnected, the assignments rather indicate tendency instead of exclusiveness.

Fourier analysis

First, a time domain signal is detected by a sample-and-hold element with a fixed sampling time interval. The application of a mathematical transform yields the transition to the frequency domain; here the dominating technique is the discrete Fourier transform (DFT) in its implementation as the fast-Fouier transform (FFT) [Lyo11].
This approach provides very high flexibility because the spectral properties can be selected after detection, within certain limits. From a microwave engineering point of view, the properties of the receiver are critical and mostly determine the capabilities of the method. As the current state-of-the-art of maximum frequency is still below 100 GHz (compare e.g. [agi13a]), frequency conversion is necessary for the application to millimeter-wave problems.
The transform of time signal segments allows adaptive frequency-time analysis. Depending on the application, for example the short-time Fourier transform (STFT), the closely related Gabor transform, wavelets or more specialized distribution methods are typically used [Coh89, KSW08].

Filter banks

Here frequency-adjacent bandpass filters select parts of the incident spectrum, which are typically detected by rectifying diodes.[3] This allows frequency-selective time-dependent analysis, and can be expressed as a hardware implementation of the STFT. The filter edge steepness and selectivity are limited by the relative bandwidth, which prohibits high frequency resolution in the upper microwave or millimeter-wave domain without frequency conversion. The total coverage of a filter bank with N channels each of the width δf is $\Delta f = N \cdot \delta f$. Practical applications are often limited by the inherent lack of flexibility and the comparatively low signal-to-noise ratio (SNR) caused by the influence of thermal noise through $P_{\mathrm{N}} = k_{\mathrm{B}} T \delta f$.[4]

[3] Compare for example [GTW91], an early implementation for gyrotron testing.
[4] k_{B}: Boltzmann constant, T: physical temperature in Kelvin

Spectrum analyzer

The standard tool for practical spectrum analysis in the microwave and millimeter-wave domain circumvents the filter bank's bandwidth and dynamic range limitation by employing a super-heterodyne arrangement with a multi-stage IF bandpass. The frequency coverage is provided by sweeping the LO, which normally follows a linear ramp. This arrangement allows the coverage of wide frequency ranges (*span*) of multiple tens of GHz, while maintaining the possibility of resolving frequencies down to the order of Hertz (see e.g. [rs213], [agi13b]).
These modern spectrum analyzers operate up to 40–67 GHz; above, external mixers are needed to extend the frequency range as described in Section 1.3.1. Few exceptions can be found; for example, Elva-1 offers a spectrum analyzer covering 120–180 GHz which uses a backward-wave oscillator in the same frequency range as the LO [elv00].
The main limitation with respect to the challenges listed before is that during the sweep time t_{sweep}, only a part of the entire span is monitored. Thus the sweep technique inherently introduces a problem in the analysis of time-dependent signals, since it connects the frequency axis with the time axis. From this also rises the rule

$$\delta f > \sqrt{\frac{\Delta f}{t_{\mathrm{sweep}}}} \tag{1.2}$$

for the minimum dynamic resolution bandwidth [Eng84, TWK98]. For a frequency resolution $\delta f = 1\,\mathrm{MHz}$ and a span of $\Delta f = 5\,\mathrm{GHz}$, this relation yields a minimum sweep time of $t > \Delta f / \delta f^2 = 5\,\mathrm{ms}$; for shorter pulses, the effective span is reduced.

Counter based

An approach especially suited for clean and frequency-agile signals is the measurement with frequency counters. Instead of a constant sampling rate, here an event-based representation derived from the signal fluctuations is employed [TWK98]. This reduces the amount of data, allows the easy analysis of phase-frequency relationships and also

enables statistical analysis. The typical measurement type is frequency vs. time, also referred to as *modulation domain analysis*. However, more complex analyses like histogram display or Allen-variance noise characterization are also directly enabled.
Older devices utilizing so-called zero-dead time counters were already capable of covering a few GHz [hp200], but modern modular systems developed with a focus on radar system characterization [agi08] feature a bandwidth of up to 18 GHz. Unfortunately, the dynamic range of such systems is normally less than 30 dB, which makes them of limited use for high-sensitivity tasks such as parasitic-oscillation tracking in gyrotrons. A second drawback results from the inherent problems which are imposed on counter-based techniques by noiselike or multi-signal spectra.

Further techniques

Of course more techniques than the ones presented here exist, such as quasi-optical approaches with scanning Michelson interferometers [AEDJ97] used for plasma diagnostics, or dispersive-line systems for analyzing ultra-short pulses [BSS07]. However, these would fall short of efficiently dealing with the challenges listed in Section 1.2, and are therefore not discussed.

1.4 Scope of this work

1.4.1 Measurement system

The goal to cover the challenges set up in Section 1.2 as generally and as efficiently as possible determines many required properties of the measurement system almost automatically. By the requirement for high frequency resolution combined with time-dependent observation of concurrent signals, maintaining a maximum dynamic range, all sweep and counter based techniques are ruled out. Consequently, a time domain approach is most suitable, which includes a subsequent transition into the spectral domain through STFT.[5] As the maximum relevant frequency of 170 GHz is out of the range of state-of the art A/D (analog-to-digital) converters, at the front-end a conversion to an IF range must be performed. The introduction of harmonic mixers for this purpose however initially violates the condition for unambiguous RF identification. In the implemented system, a novel dual-channel technique is employed, which facilitates instantaneous mixing harmonic identification in combination with separation of the HSI and LSI bands for a reconstruction of the true RF spectrum. in the following, the entirety of the system is referred to as the pulse spectrum analysis (PSA) system, and is described in Chapter 4.

1.4.2 Investigations

The second part of the work is dedicated to the application of the measurement system. Assorted primary results demonstrate the capabilities of the PSA system in Chapter 5, including brief discussions and analyses of the documented spectral phenomena.
Special attention is given to parasitic oscillations, which were one prime motivation for the development of the PSA system. This topic, as well as the investigation of the tuning effects at the start of longer gyrotron pulses, are treated in the Chapters 6 and 7.
One important goal pursued throughout the work is the complementary application of theory and simulation. Only through the rigorous fusion of these domains, can a maximum of knowledge be gained. This motivates the preparative Chapter 3, which puts the possibilities and limitations of both measurement and simulation into context.

[5] An extremely sophisticated measurement system for the application of electron-cyclotron emission spectroscopy having comparable properties is presented in [TBW+09]. It consists of a fundamental mixer pumped with 140 GHz, which divides the D-band (110–170 GHz) in two sub-bands to be sampled by ultra-fast A/D-converters. Despite the high data load, it is capable of real time operation due to custom-developed FPGA (field-programmable gate array) based data processing. While it would be interesting to incorporate such a system into a gyrotron test stand, it fits better to the requirements and budgetary capabilities of a fusion reactor.

2 Gyrotron setup and behavior

In the following, a technological overview of the gyrotron is given, along with physical quantities typically used for its characterization. For this, first the electron-optical system is described.
More extensive information on the gyrotron can be found in the literature, see for example [Nus04, KBT04, Edg93] or [NTP14, Chu04], the latter references also containing a historical perspective. A short introduction as well as a comprehensive overview on the state of the art in gyro-devices is given in [Thu14].
A schematic overview of a high-power gyrotron along with the magnetic flux density along the z-axis is given in Figure 2.1.[1]

2.1 Electron-optical system

The gyrotron cathode contains an emitter ring for the generation of an annular electron beam. Electrons leaving the emitter are accelerated by the potential difference between cathode and anode, V_{acc}, towards the cavity and eventually the collector, while being guided by a strong static magnetic field created by an external superconducting magnet. The combination of the external magnetic field lines and the applied electrostatic potential, shaped by the cathode area and overall internal geometry of the gyrotron, form the electron-optical system.

2.1.1 Electron beam and macroscopic properties

In the typical arrangement, the static electric fields at the emitter ring surface are approximately perpendicular to the magnetic field lines. This design, the magnetron injection gun, ensures that the emitted

[1] It should be noted that the high voltage power wiring in Figure 2.1 is highly simplistic; a more accurate description is available in [DAF+03].

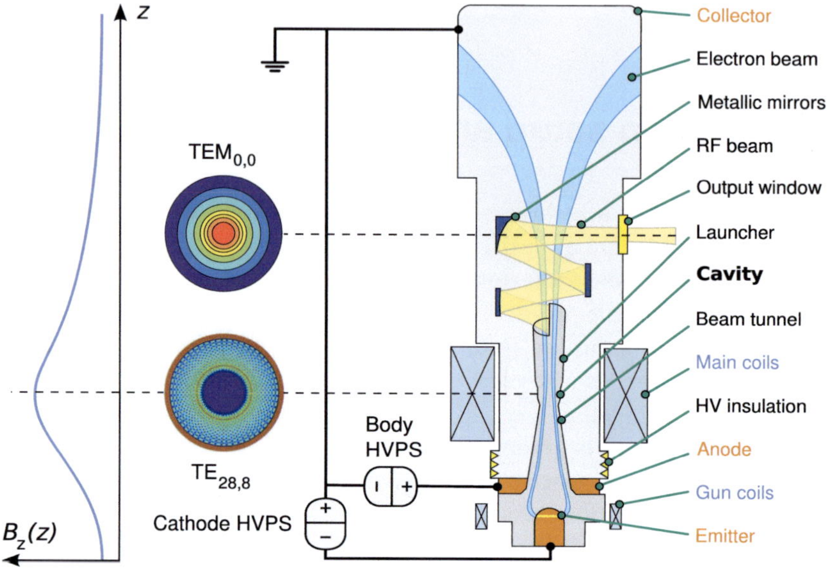

Figure 2.1: Modern high power cylindrical-cavity gyrotron with basic on-axis magnetic flux density and schematic high voltage power supplies (HVPS)

electrons have a precisely controllable configuration of the transversal and axial velocity components $v_\perp$ resp. $v_\|$ in the gyrotron's cylindrical coordinate system. The most common description of this configuration is the pitch factor

$$\alpha = \frac{v_\perp}{v_\|} \quad , \tag{2.1}$$

which varies with the magnetic flux density along the beam trajectory toward the collector.
The Lorentz force $\vec{F} = -e(\vec{v} \times \vec{B})$ resulting from this arrangement causes the focused electrons to follow helical trajectories around magnetically determined guiding centers. Macroscopically, this is characterized by the beam radius r_b, which is the mean distance of the local guiding center from the z-axis. The distance between the actual spiral

trajectory and the guiding center is given by the Larmor radius

$$r_{\mathrm{L}} = \frac{p_{\perp}}{|q| B_{\mathrm{z}}} \quad , \tag{2.2}$$

which also defines the approximate minimum beam thickness $2r_{\mathrm{L}}$.
Since for fundamental harmonic operation, the generated RF is very close to the electron cyclotron frequency as defined in (1.1), multi-Tesla magnetic flux densities are necessary for millimeter-wave oscillations at the fundamental ECR frequency. Typical cathode voltages for fusion gyrotrons are 75–90 kV, which yields weakly relativistic electron velocities and corresponding particle energies of

$$W_{\mathrm{k}} = eV_{\mathrm{b}} = W_0 \left(\gamma - 1\right) \tag{2.3}$$

with the electron rest energy $W_0 = m_e c_0^2$ and the Lorentz-factor

$$\gamma = \frac{1}{\sqrt{1-\beta^2}} \quad = \quad 1 + \frac{W_{\mathrm{k}}}{W_0} \quad \approx \quad 1 + \frac{W_{\mathrm{k}}}{511\ \mathrm{keV}} \tag{2.4}$$

containing the normalized electron velocity

$$\beta = \frac{v}{c_0} = \sqrt{\beta_{||}^2 + \beta_{\perp}^2} \quad . \tag{2.5}$$

This collection of values is typically referred to as *beam parameters*, as they give the primary information on the properties of the electron beam relevant for calculations and simulations. Dropping redundancy, the entire information can be expressed in the terms of α and γ.
Secondary information lies in the statistical distribution of these parameters around their average values, where especially $\delta\alpha$ is important.[2]
A simple and regularly made assumption is that of a roughly Gaussian distribution of the parameter spread. This interprets the pitch factor as a random variable A, with a probability density function $f_A(\alpha)$ characterized by only the expectation value $E[A] = \mu = \overline{\alpha}$ and the standard deviation σ.

[2]Depending on the case, the separated velocity spreads $\delta\beta_{||}$, $\delta\beta_{\perp}$ are also employed.

The "RMS-spread" (root-mean-square) is then

$$\delta\alpha_\sigma = \frac{\sigma}{\mu} \quad , \tag{2.6}$$

i.e. the standard deviation normalized with the expectation value.
An alternative and more abstract definition, the "relative spread", is employed for example in [Tsi01]. For this, the difference of the 90% and 10% quantiles[3] is normalized to the expectation value μ as

$$\delta\alpha_{\mathrm{rel}} = \frac{q_{\alpha,0.9} - q_{\alpha,0.1}}{\mu} \quad . \tag{2.7}$$

The advantage of the latter formalism is the lack of inherent statistical assumptions, allowing it to retain its meaning in the presence of arbitrary spread distributions.
For Gaussian distributions a fixed relationship between $\delta\alpha_\sigma$ and $\delta\alpha_{\mathrm{rel}}$ can be found:

$$\delta\alpha_{\mathrm{rel}} = \frac{2.5631\sigma}{\mu} \quad \Rightarrow \quad \frac{\delta\alpha_{\mathrm{rel}}}{\delta\alpha_\sigma} = 2.5631 \quad . \tag{2.8}$$

Note that in the case of asymmetric distributions, as easily introduced by beam-wave interactions, the expectation, mean and most probable value of a parameter are no longer identical, so that the normalization reference in (2.6) and (2.7) can become disputable.

2.1.2 Adiabatic quantities

The magnetic flux density in coil arrangements as they are used in the gyrotron field varies relatively smoothly along the z-axis, which allows the application of adiabatic invariants following from Busch's theorem [Rei08] with two important consequences:

[3] An x-quantile q_x of a random variable A with a probability density function $f_A(\alpha)$ is implicitly defined as the value of α for which the expectation value equals x, i.e. $E[A \leq q_x] = \int_{-\infty}^{q_x} f_A(\alpha)\,\mathrm{d}\alpha = x$.

1. The electron trajectory guiding centers coincide with the magnetic field lines. This connects the axial variation of beam radius and magnetic field as

$$\frac{r_\mathrm{b}(z_2)}{r_\mathrm{b}(z_1)} \approx \sqrt{\frac{B_\mathrm{z}(z_1)}{B_\mathrm{z}(z_2)}} \quad ; \tag{2.9}$$

   the term $b = B_\mathrm{z}(z_1)/B_\mathrm{z}(z_2)$ is the magnetic compression ratio.

2. Since the magnetic field transfers kinetic electron energy between the axial and the transverse plane, α changes with $B_\mathrm{z}(z)$ in a likewise way as r_b does. Considering the relativistic transverse momentum z-translation to be constant[4] , from

$$\gamma m_\mathrm{e} v_\perp \approx const \quad \Rightarrow \quad \frac{\beta_\perp(z_2)}{\beta_\perp(z_1)} \approx \frac{\gamma(z_1)}{\gamma(z_2)} \sqrt{\frac{B_\mathrm{z}(z_2)}{B_\mathrm{z}(z_1)}} \tag{2.10}$$

   follows that $\alpha(z)$ increases with $B_\mathrm{z}(z)$.

2.1.3 Electrostatic voltage depression effect

The electron kinetic energy is one determining parameter of the gyrotron interaction. However, the electron beam itself constitutes a significant amount of negative charge inside the gyrotron body.[5] In combination with the external potential of the metallic geometry itself, this imposes a repelling potential on electrons which emerge at the emitter surface. The consequence is a decrease ("depression") of the z-local effective beam voltage $V_\mathrm{b} = W_\mathrm{kin}/e$ with respect to the technical cathode voltage V_cath. This quantity

$$V_\mathrm{dep} = V_\mathrm{cath} - V_\mathrm{b} \tag{2.11}$$

is generally referred to as the *depression voltage*[6] [KBT04].

---

[4] I.e. the "first adiabatic invariant" in plasma physics.

[5] This is often called "space charge".

[6] This term can be ambiguous in the context of depressed collector operation. Here V_body is a positive voltage with respect to the collector for efficiency enhancement. In this work, the *voltage depression* V_dep refers to the space-charge related effect, not a technical body-collector potential depression.

For azimuthally symmetric annular electron beams placed coaxially in a cylindrical metallic geometry, this voltage has been calculated analytically[7] [DK81] to

$$V_{\mathrm{dep}} = \frac{I_{\mathrm{b}}}{2\pi\epsilon_0 v_{\|}} \ln\left(\frac{r_{\mathrm{w}}}{r_{\mathrm{b}}}\right) \quad . \tag{2.12}$$

For the W7-X gyrotron cavity with $r_{\mathrm{c}} = 20.48\,\mathrm{mm}$, at a nominal operating point with $V_{\mathrm{cath}} = 80\,\mathrm{kV}$ and $r_{\mathrm{b}} = 10.1\,\mathrm{mm}$, (2.12) yields $V_{\mathrm{dep}} \approx 5.4\,\mathrm{kV}$. Thus, at the location of interaction the particles have an effective $W_{\mathrm{k}} \approx 74.6\,\mathrm{keV}$, which is $\sim 6.7\%$ below $e \cdot V_{\mathrm{cath}}$.[8]

2.2 Microwave fields and gyrotron interaction

In the following, the basic characteristics of the gyrotron interaction are outlined, as far as suitable for the problems discussed in the work. More elaborate and detailed descriptions can be found e.g. in [KBT04, Nus04, Edg93], or in the literature on the gyrotron interaction simulation codes listed in Section 3.2.2.

2.2.1 Waveguide modes

The gyrotron resonator, or cavity, is located at the peak position of the static external magnetic field, and normally contains a cylindrical waveguide section. A system of eigensolutions for the electromagnetic fields in such a geometry is given by the transverse-electric (TE) and transverse-magnetic (TM) mode formalism [Har01].
Macroscopically, these waves can be characterized by the dispersion relation

$$k_0^2 = k_{\|}^2 + k_{\perp}^2 \tag{2.13}$$

[7]It must be noted that the derivation of this expression assumes locally constant geometry and beam radii and are only suitable for "smooth" radius changes.

[8]As visible in (2.12), this effect concerns only the axial electron velocity. It can not be emulated or corrected by simply varying the cathode voltage.

with the free-space propagation constant

$$k_0 = \frac{2\pi f}{c_0} \quad . \tag{2.14}$$

The transverse phase number

$$k_\perp = \frac{\chi_{m,n}}{r_\mathrm{w}} \tag{2.15}$$

defines the possibility of propagation by including the influence of the waveguide radius r_w and the mode eigenvalue $\chi_{m,n}$. For a TE wave, $\chi_{m,n}$ is the nth zero in $J'_m(\chi)$, which is the derivative of the mth order Bessel function of the first kind.
This yields a hyperbolic dispersion relation:

$$f = \frac{c_0}{2\pi}\sqrt{k_\parallel^2 + \left(\frac{\chi_{m,n}}{r_\mathrm{w}}\right)^2} \quad . \tag{2.16}$$

A special case is $k_\parallel = 0$, yielding with

$$f_\mathrm{cut} = \frac{c_0 \chi_{m,n}}{2\pi r_\mathrm{w}} \tag{2.17}$$

the cut-off frequency, below which the mode is subject to aperiodical attenuation. Only modes for which the given combination of f, $\chi_{m,n}$ and r_w in (2.16) yields a real $k_\parallel > 0$ are "above cut-off" and able to propagate through the waveguide.

2.2.2 Interaction

In the gyrotron interaction, a TE waveguide mode phase-synchronizes with the electron beam over a finite interaction length, which leads to the possibility of energy exchange between beam and wave. The transport of energy from the beam to the wave here means extraction of kinetic energy from the electrons; this makes an electric field term which decelerates the electrons the "acting component" of the wave.
A peculiarity of the gyrotron interaction is that it describes the interaction of the electron beam with a TE waveguide mode very close to

cut-off. This has a number of inherent consequences, e.g. that the wave has a high phase velocity, coining the term "fast-wave" interaction for gyro-devices. More important, however, the purely transverse field structure of TE waves also causes the energy extraction to affect only the transverse velocity component $\beta_\perp$ of the electrons.
The prerequisites of an efficient interaction are apart from others thus defined by the radial dependence of the azimuthal field component E_ϕ, which for an ideal $\mathrm{TE}_{m,n}$ eigenmode of a cylindrical cavity is dominated by the Bessel term $J'_m(k_\| r)$. A basic measure for the possible coupling between electron beam and waveguide mode field can be derived from such considerations; for fundamental-harmonic gyrotron operation, this *coupling coefficient* as defined in [Ker96, KBT04] equals

$$Ge_{m,n} = \frac{J_m(k_{\perp m,n} r_\mathrm{b})}{J_m(\chi_{m,n})\sqrt{\pi(\chi^2_{m,n} - m^2)}} \quad . \tag{2.18}$$

Cylindrical waveguide modes can rotate in azimuthal direction, giving rise to the separation into co- and counter-rotating modes with respect to the electron rotation direction. Within this work, explicitly co-rotating modes are denoted with a "$-$"-, and counter-rotating modes with a "$+$"-superscript.

2.2.3 Basic beam-wave synchronization

Simple resonance condition

From the parameters given, a simplified resonance condition can be derived [Chu04]:

$$\omega - k_\| v_\| \gtrsim \frac{\Omega_0}{\gamma} \tag{2.19}$$

It expresses the synchronicity of a locally plain wave's phase front with the beam electrons. Traveling along the z axis with $v_\|$, these gyrate with $\Omega_\mathrm{r} = \Omega_0/\gamma$ (compare eq. 1.1), while the wave is characterized by the propagation term $\omega t - k_\| z$. The RF field remains in phase with the beam electrons long enough to allow relevant interaction only when this condition is fulfilled.

Depending on their relative phase position, electrons are either accelerated or decelerated by the RF electric field. A net energy exchange is only possible through *phase focusing*, or *bunching*; it depends on the Lorentz factor in Ω_r, connecting the energy of the electrons with their gyrating phase. Electrons which are decelerated lose energy, increasing their Ω_r, while for accelerated electrons the opposite holds.
For a stable net energy transfer from the beam to the RF field therefore $\omega \gtrsim \Omega_\mathrm{r}$ is necessary: An electron which loses energy exhibits increased Ω_r, therefore remaining in the favorable energy-extracting phase relative to the RF wave longer. On the contrary, electrons which are gaining energy reduce their residence time in this phase position, thus preventing effective energy extraction from the wave. This mechanism tends to continuously accumulate the major fraction of the electrons in a similar phase position with respect to the RF wave, hence the term "bunching".
The resonance condition between a waveguide mode following (2.13) with the electron beam can be illustrated in the Brillouin diagram, as done e.g. in [Thu14, Chu04, Edg93]. For a cylindrical waveguide mode, the hyperbolae in Figure 2.2 define allowed propagation solutions obtained from (2.16), while the straight beam lines contain the information on the macroscopic electron beam parameters by

$$\omega = v_\parallel k_\parallel + \Omega_\mathrm{r} \quad . \tag{2.20}$$

Intersection and boundary points of the hyperbolas with beam lines

$$\frac{\omega}{c_0} = \sqrt{\left(\frac{\omega - \Omega_\mathrm{r}}{v_\parallel}\right)^2 + \left(\frac{\chi_{m,n}}{r_\mathrm{c}}\right)^2} \tag{2.21}$$

denote configurations which fulfill the resonance condition between beam and wave in the cavity ($r_\mathrm{w} := r_\mathrm{c}$ in eq. 2.15). In the figure, the dispersion curves of the three lowest-order TE circular waveguide modes $\mathrm{TE}_{1,1}$, $\mathrm{TE}_{2,1}$ and $\mathrm{TE}_{0,1}$ are plotted together with a sample beam line. For $k_\parallel = 0$, the beam line angular frequency equals Ω_r.
Point ⓐ marks a typical gyrotron interaction point: The beam line is a tangent to the hyperbola in the forward wave domain ($k_\parallel > 0$), and the interaction frequency ω_a is close to the mode cut-off frequency $\omega_{\mathrm{cut}2,1}$.

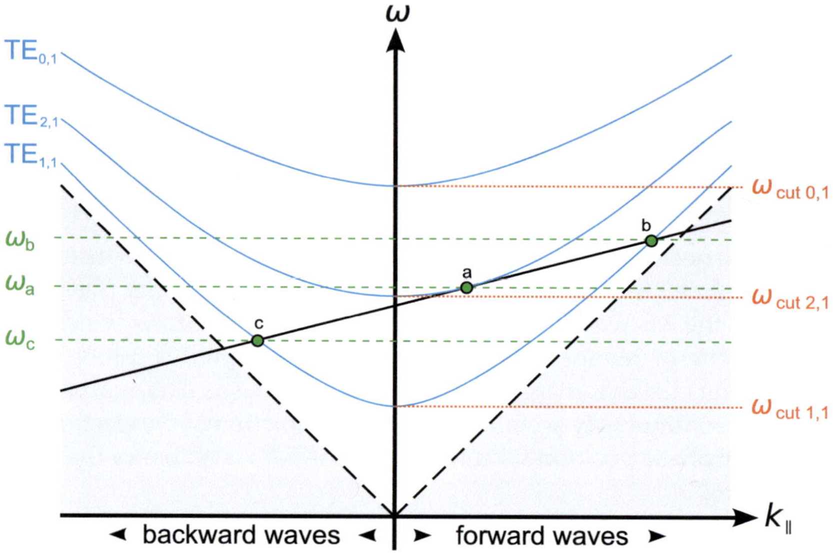

Figure 2.2: Sample circular waveguide mode dispersion diagram (Brillouin diagram) with electron beam line, speed-of-light-lines (dashed), cut-off frequencies and marked resonance points (qualitative)

The intersection ⓑ is further away from cut-off, but also in the forward wave range. Note that even though $\omega_\mathrm{b} > \omega_\mathrm{a}$, the mode of interaction is of lower order, being the fundamental. Backward waves can also be excited, as illustrated by point ⓒ. The RF frequency ω_c is lower than Ω_r only for these waves.

Limitations

The formulation introduced above describes the resonance between an unperturbed mono-energetic electron beam and an ideal loss-less waveguide mode in a uniform geometry of infinite length. In addition to these obvious idealizations, further limitations exist.

Even in the cavity with its cylindrical walls, the mode pattern is not purely sinusoidal in the axial direction, since the gyrotron cavity length is of the order $\pi/k_{\|}$. This causes a deviation of the waveguide mode's field structure from ideal treatment, and therefore also changes its dispersion characteristics. The presence of the electron beam in the geometry has a similar, but potentially even more profound effect, since it imposes a modification of the boundary conditions for the mode.
Furthermore, a stationary gyrotron oscillation does not simply take place with the beam as it enters the cavity, it rather is a dynamic interaction process in which the beam parameters are modified. This is not taken into account by (2.21); however, if the RF power generated is known, a coarse macroscopic estimation can be proposed: Assuming that P_{RF} is extracted from all electrons uniformly, the average effective beam voltage after the interaction can be formulated as

$$V_{\mathrm{b}}' \approx V_{\mathrm{b}} - \frac{P_{\mathrm{gen}}}{I_{\mathrm{b}}} \quad . \tag{2.22}$$

This "spent beam" energy automatically defines a modified γ' over (2.3), which for $\beta_{\|} = \beta_{\|}'$ yields a new pitch factor

$$\alpha' = \frac{\sqrt{1 - (1/\gamma')^2 - \beta_{\|}^2}}{\beta_{\|}} \quad . \tag{2.23}$$

The values most appropriate for the application of resonance conditions might lie between the "fresh" and the "spent" beam parameters; this is explored in Section 3.3.1.
Another point is raised casually in [Ker96], where the statement is made that the wave frequency ω as defined in (2.21) only marks the lower edge of an excitation band which approximately has the width of the inverse residence time of an electron in the interaction zone.

2.3 RF-relevant geometry

In the following, the parts of the gyrotron which are relevant for the generation and guidance of RF energy are introduced. An illustrative schematic of the typical waveguide geometry is given in Figure 2.3.

2.3.1 Gyrotron cavity

As mentioned before, the cavity is the structure in which the nominal gyrotron interaction takes place. The midsection consists of a cylindrical resonator with a precisely defined radius. Its quality factor is strongly influenced by its axial extent and the non-linear tapers on both ends, which makes them the subject of careful optimization. Since the stationary fields of a waveguide mode operating in the midsection protrude into these adjacent sections, typically the combined entity of downtaper, cylindrical section and uptaper are referred to as the cavity.

Together with the discrete eigenvalue $\chi_{m,n}$ of a $\mathrm{TE}_{m,n}$ mode, the cavity wall radius r_c is one of the determining factors of the final oscillation frequency. As the wall radius of the cylindrical cavity midsection is the minimum radius the nominal cavity mode encounters, here also the highest RF-induced wall losses occur. For megawatt class gyrotrons, these losses can easily reach $2\ \mathrm{kW/cm^2}$.

Toward the electron gun, the geometry radius decrease of the downtaper section forms a reflective boundary through cut-off for modes which are excited at low $k_\|$ in the cavity, as is normally the case for gyrotron-type interactions.

2.3.2 Cathode to cavity

In the area between the cathode and the cavity downtaper section, the *electron beam compression zone* or *beam tunnel*, the beam radius reduces adiabatically (compare Section 2.1.2). Since the high-energetic beam can very easily excite fields in this area, the task of the beam tunnel is to prevent beam-wave interactions prior to the cavity; however, unwanted oscillations are a well-known problem.

The beam tunnel is normally implemented as an irregular structure, in order to impede electromagnetic waveguide mode propagation and excitation as strongly as possible. Typical solutions include pseudo-chaotic stainless-steel structures [LST11], absorbing full-ceramic walls [SSH+01, RGI+13], or stacked alternating arrangements of corrugated diffractive copper rings and absorbing ceramic rings [TG10, GDF+10, BFB+04]. The steep constriction between downtaper and beam tunnel is often referred to as the *cut-off section*.

2.3.3 Cavity to output window

Downstream of the cavity, the wall radius increases in a non-linear uptaper section. The radius variation necessarily introduces partial reflections of the wave propagating from the cavity, forming a continuous second boundary of the actual desired interaction region.
In modern high-power gyrotrons, the uptaper section is followed by a quasi-optical mode converting antenna, the launcher. Its task is the conversion of the high order $TE_{m,n}$ cavity mode into a mode suitable for external transmission lines, which are extensively covered in [TK02]. For high power applications there exist only two basic transmission line types: The first is quasi-optical and therefore requires

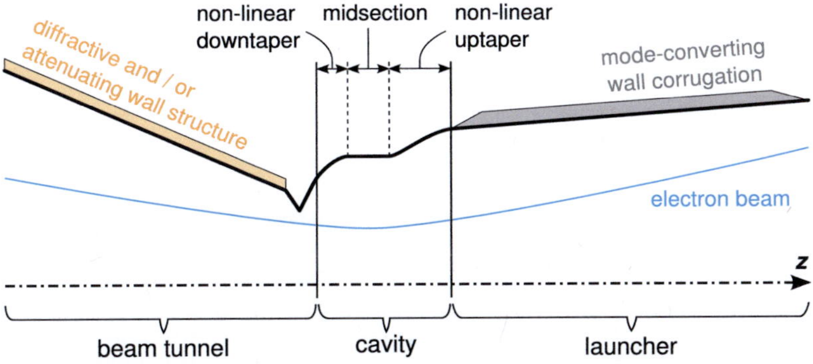

Figure 2.3: Schematic longitudinal cut through waveguide geometry

a Gaussian fundamental mode, $TEM_{0,0}$, and is employed at the KIT high power gyrotron test stand and the W7-X ECRH facility in Greifswald. The other solution is the use of corrugated overmoded circular waveguides, which use the $HE_{1,1}$ fundamental mode. This hybrid mode is different from $TEM_{0,0}$, but closely related, which has lead to virtually all gyrotron quasi-optical launchers being designed for $TEM_{0,0}$.
The launcher consists of a slightly conical waveguide section. While propagating through the launcher, a mode is gradually converted by specific wall corrugations of sub-wavelength depth. These allow directed manipulations of the waveguide mode's amplitude and phase, but must be tailored to a designated design mode to have maximum effect. A cut at the launcher end leads to radiation of the converted linearly polarized mode in transverse direction, where it is directed at the microwave output window by up to four metallic mirrors. The window acts as a vacuum barrier, while it must present minimum reflection as well as minimum absorption to the beam. Chemical-vapor deposited synthetic diamond is the state-of-the-art material, which is normally employed as a single disc, using a $\lambda/4$-resonance [Thu01]. Broadband setups, on the other hand, typically feature arrangements using the Brewster-angle [YBD+03].

2.4 Behavioral aspects

In the following, a summary of additional details on the general operation behavior of gyrotrons which are not obvious from the above explanations is provided.

2.4.1 Mode competition and starting currents

A strongly interacting waveguide mode dynamically arranges the beam electrons in bunches corresponding to its azimuthal index and oscillation frequency. In a first-order treatment, this automatically tends to suppress other modes in the geometrical range of interaction, so the basic scenario is characterized by one dominant mode [ZZM06].

It can be assumed that at many operating points when another mode rises, there is a short meta-stable period of competition, which is ended by the suppression of one of the two modes. Of course it is also possible that the competition is prolonged, resulting in perturbed multi-frequency operation.
In contrast to the highly simplified and qualitative description above, non-linear theoretical investigations [Nus81] and self-consistent numerical calculations [ZZM06] show the possibility of stationary multi-mode activity.[9] Typically the involved modes are coupled, and share specific relationships; e.g. for certain operating points, several competitors which share a common mode index number are known to be likely to start together [APPV07, SNA12]. Also in recent KIT experiments with a 2nd harmonic 28 GHz $TE_{1,2}$ gyrotron [MIP$^+$13], the parallel cavity operation of the nominal mode with a strong backward wave at the fundamental frequency of 14 GHz with $TE_{1,1}$ was observed routinely at certain operating points [Mal14].
Further prominent competitors with a given nominal mode are the direct azimuthal neighbors, and all modes sharing the radial index, as they match the azimuthal bunching. Further notorious modes are $TE_{m-2,n+1}$ and $TE_{m-3,n+1}$ with respect to the nominal mode [Ker96, Ber11]. The latter mode also lies within few hundred MHz of the nominal mode, which can cause ambiguity in measurements.
Modes in gyrotron cavities can always be assigned a minimum beam current necessary for starting from the noise level, which is related to their coupling strength to the electron beam and their interaction efficiency [BJ88]. At an arbitrary operating point, the mode with the lowest starting current is typically expected to start first. Also, sometimes one mode starts, and is later suppressed by another one as parameters change; this is typical for start-up scenarios.
It is important to note that these mechanisms introduce a significant hysteresis of the gyrotron over its external parameters: If the cavity mode is switched due to a temporary shift of a parameter, even after

[9] Note that in mode competition studies often all waveguide modes except the desired operating mode are termed parasitic, conflicting with the nomenclature used in this work (compare Sections 1.2.5 and 6.1.1).

the reset of this parameter the gyrotron is likely to be in a different state than initially. This has been demonstrated and diligently investigated in [SKT+07].
The macroscopic interaction description of (2.2) assumed that an RF field is already present, which interacts with the electron beam. This viewpoint is justified because of the helical electron trajectories which necessarily indicate constant centripetal acceleration, causing the continuous emission of bremsstrahlung around the z-local Ω_r. Since all electromagnetic radiation must obey the boundary conditions in the waveguide, this means that one can expect all modes with frequencies around Ω_r which are able to propagate to be actually present, forming a background spectrum.

2.4.2 Schottky-effect

Though the gyrotron's electron gun is designed for thermionic emission, the electric fields on the emitter ring influence emissivity through a skew of the material work function. This range of operation is often called Schottky-emission, and can be covered analytically by a modification of the Richardson-Dushman equation, which in its basic form only describes purely thermionic emission [MGJ56].
The relevant effect in the experimental situation is that the beam current follows the cathode voltage, which happens much faster than the response to heating because of the gun's considerable thermal inertia.

2.4.3 Pulse characterization from duration

Both from external technical necessities and internal physical phenomena, experimental gyrotron operation can be reasonably divided into short- and long pulse operation.
Short-pulse (SP) operation is constrained to pulse durations up to a few ms. This allows the use of a comparatively simple capacitor-based power supply and a ballistic calorimeter mounted directly at the output window. Since even for megawatt pulses the DC power does not exceed a few kJ, this type of operation is relatively safe and suitable for

a wide range of experiments. Due to the typical pulse repetition times below one second, most components can remain passively cooled.

In long-pulse (LP) operation, the pulses take seconds to hours. This requires a major transformer-based power supply, capable of delivering multi-megawatt CW power. The requirements for the RF load and the cooling rise in the same way: A dedicated LP load and powerful water-cooling for all RF-facing components are mandatory.

During the first seconds of a long pulse, multiple internal effects modify the gyrotron interaction operating point, leading to transient behavior in frequency and instantaneous power. In Figure 2.4, the idealized typical frequency shift versus pulse time index is displayed and connected with the defined ranges for SP and LP operation.

SP operation is within a quasi-stationary regime; from experiments with W7-X gyrotrons, the frequency shift in these first milliseconds can be estimated to be below 1 MHz/ms. After a transitory regime, roughly in the range 0.01–2 s, the final LP operating point is reached,

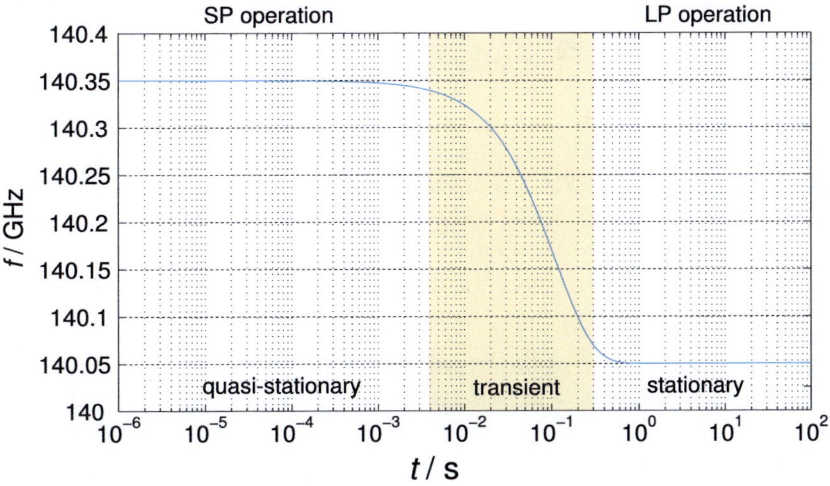

Figure 2.4: Illustration of exponential frequency development versus logarithmic pulse time

which is considered a truly stationary condition concerning the internal physical phenomena. The changes in the transition regime can be attributed to two independent physical effects inherent to LP operation, which are briefly explained in the following.

Thermal expansion

The loss power P_{diss} deposited in the cavity walls is of the order of up to $2\,\mathrm{kW/cm^2}$ in usual operation. Even using an extremely sophisticated cooling system, the physical temperature of the cavity and its surroundings will heat up by more than $100\,^\circ\mathrm{C}$. The resulting thermal expansion of the wall radius r_{w} can be expressed in terms of the linear expansion coefficient a_{th} in

$$\frac{\Delta L}{L} = a_{\mathrm{th}} \Delta T \quad , \tag{2.24}$$

where ΔL is the expected change of the length L over the temperature difference ΔT. Already in the simple resonance condition (2.21) a direct influence on the oscillation frequency can be deduced.

Voltage depression neutralization

As laid out in Section 2.1.3, the space charge by the electron beam lowers the local electron energy available for the gyrotron interaction. Even though gyrotrons are operated with ultra-high internal vacua of $p < 10^{-7}\,\mathrm{mbar}$, there is enough residual gas inside the gyrotron body to be gradually ionized by the electron beam. This process can be generally described over the ionization rate

$$g_{\mathrm{i}} = \frac{\text{number of ionizations}}{\text{time interval}} \quad , \tag{2.25}$$

which in combination with a known neutral particle density n_{g} defines a total ion creation rate. As the created ions agglomerate along the static negative potential of the beam, the voltage depression effect is gradually neutralized. This in principle can increase the effective kinetic energy by several kV (compare Section 2.1.3), which is a relative change of several percent.

2.4.4 Long-line effect

This effect describes the behavior of a free-running oscillator facing power reflection from an electrically distant location. A basic consequence is dynamic load-pulling of the oscillation frequency through the delayed power feed-back, resulting in a dynamically altered oscillation frequency and power level [Pri56]. It has been already observed in gyrotrons experimentally and theoretically before, as laid out in e.g. [MTT+90, AJCN92, BGJ+92]. Depending on the quality factor of the combination of resonator and transmission line, the strength of the reflection and its distance, this effect can cause coherent modulation in the form of a line spectrum, incoherent wide band modulation or discrete frequency jumping.
For coherent modulation of the oscillation with f_{M}, the time constant $\tau_{\mathrm{M}} = 1/f_{\mathrm{M}}$ of this periodic process can be related to the round-trip time delay τ_{RT} of the RF wave on the line between the oscillation and the reflection, governed by the line length L and the propagation speed v_{en}, in the form

$$\tau_{\mathrm{M}} \sim \tau_{\mathrm{RT}} = \frac{2L}{v_{\mathrm{en}}} \quad . \tag{2.26}$$

It is logical to use this equation to relate observed modulation sideband offsets to the probable distance of the responsible reflection position, in crude analogy to time-domain reflectometry.
Figure 2.5 contains the approximate round trip times τ_{RT} of the beam-guiding elements versus the geometrical distance along the RF path from the end of the interaction region for W7-X gyrotron long pulse operation at KIT. Also the corresponding characteristic long-line frequencies f_{M} are plotted for easy reference. The data were calculated for the nominal W7-X gyrotron mode $\mathrm{TE}_{28,8}$ at 140.1 GHz, with the end of the cylindrical cavity section as the reference point $z = 0$ mm. During quasi-optical transmission, i.e. along the whole RF beam path after exiting the launcher, the energy velocity v_{en} is equal to the speed of light, so in this range uniform behavior regardless of frequency and cavity mode can be expected. Between the end of the cylindrical resonator and the launcher end, however, the waveguide dispersion

must be considered, which causes a radius dependent group velocity $v_{\mathrm{en}} = v_{\mathrm{G}}(z) < c_0$. Total delay times can be estimated through numerical integration over the group velocity term

$$v_{\mathrm{G}}(r) = \frac{\partial \omega}{\partial k_{\|}} = c_0 \sqrt{1 - (k_{\perp}/k_0)^2} \tag{2.27}$$

along the z-axis. One pair of curves in the graph includes the result of this operation, while for reference also a second pair is included which assumes $v_{\mathrm{G}} = c_0$ also for the entire path between cavity exit and launcher. This allows to judge the relevance of the dispersion.
The *factor* curve displays $f_{\mathrm{M,c0}}/f_{\mathrm{M,vG}}$, respectively $\tau_{\mathrm{RT,vG}}/\tau_{\mathrm{RT,c0}}$. A strong impact of the waveguide dispersion is only noticeable for interior reflections; for window reflections the factor between the hypothetic dispersion-free curve and the $v_{\mathrm{en}} = v_{\mathrm{G}}(z)$ is below 1.5, and converges to one for higher distances.
For an assumed reflection from the launcher end this treatment yields with $\tau_{\mathrm{RT}} \approx 4\,\mathrm{ns}$, respectively $f_{\mathrm{M}} \approx 250\,\mathrm{MHz}$, a relatively high frequency. The values for components in higher distance are soon diminished by the additional round trip delays from the rest of the transmission path, e.g. at the window $\tau_{\mathrm{RT}} \approx 10\,\mathrm{ns}$ defines a characteristic "long-line-frequency" of about $100\,\mathrm{MHz}$.
The lowest modulation frequency expected for this simple estimation would come from the reflection at the highest distance possible, the load end. Here a round-trip delay of $50\,\mathrm{ns}$ causes an expected modulation frequency of $20\,\mathrm{MHz}$.
It must be noted that although such connections seem to be obvious and the long-line effect, as established in multiple sources (e.g. [AJCN92, GZ98, GS03]), constitutes a straightforward approach, the data gathered from relation (2.26) might be a rather crude indicator. The few direct comparisons of measured modulation frequencies with dedicated simulations in literature find qualitatively matching spectra, but frequencies which are merely of a similar order of magnitude as the values predicted by the long-line considerations [AJCN92, GS03].

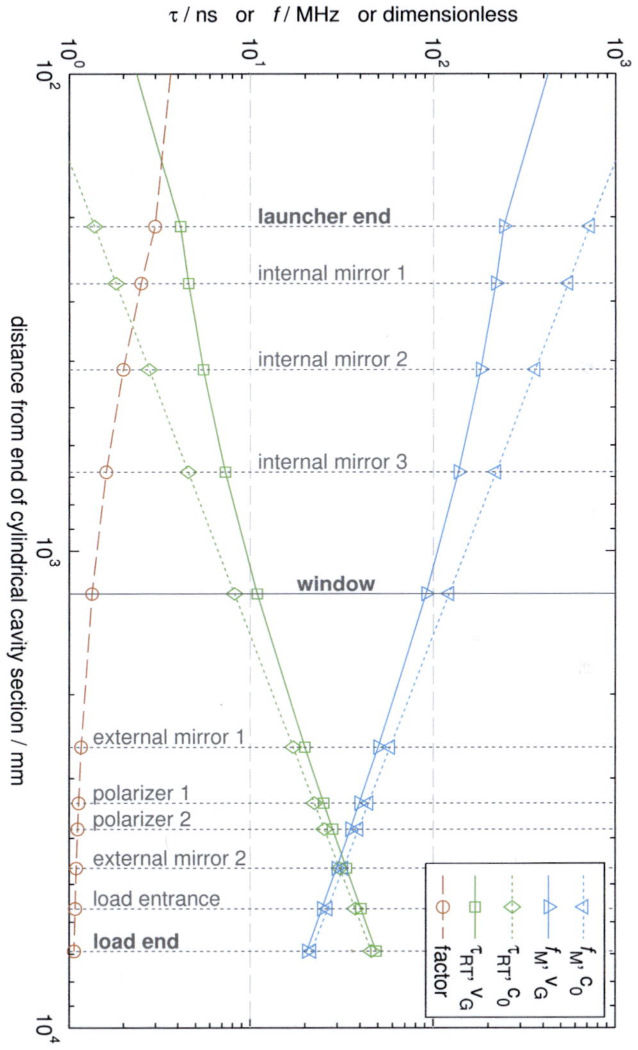

Figure 2.5: Round-trip times τ_{RT} and corresponding characteristic frequencies f_{M} for $v = c_0$ and $v = v_{\mathrm{G}}$ treatment, and the factor between both solutions

3 Measurement and simulation

3.1 Gyrotron experiments

This Chapter provides a condensed summary of high power gyrotron investigation by means of experiment and simulation. After a review of the experimental conditions and quantities, the simulation techniques and tools are presented. In order to establish and qualify a link between the two approaches, the Chapter leads to a rigorous comparison of the methods introduced in a cathode voltage sweep.
Especially the basic uncertainties and the inherent limitations of the practical as well as the theoretical methods are emphasized.

3.1.1 Experimental settings

The key parameters which define an individual operating point for a given gyrotron setup are the superconducting-magnet coil currents which govern the electron beam focusing and interaction, the cathode voltage $V_{\mathrm{cath}} < 0$, the beam current I_{b}, and in the case of depressed collector operation the body voltage $V_{\mathrm{body}} > 0$. The total acceleration voltage hence is

$$V_{\mathrm{acc}} = V_{\mathrm{cath}} - V_{\mathrm{body}} < 0 \quad . \tag{3.1}$$

However, in the rest of this work only absolute voltage values will be used, dropping the signs.
At pulse start, the voltages and the beam current are ramped by the HVPS to the set operating point within few hundred microseconds, yielding slew rates around $250\,\mathrm{kV/ms}$. During this process, the beam current follows the cathode voltage as governed by the Schottky emission effect (see Section 2.4.2).
Uncertainties are mostly introduced by the high absolute values of the parameters themselves. The voltages must be monitored over high

voltage dividers, which have a specified accuracy of 1%. For a cathode voltage of $80\,\mathrm{kV}$, this already introduces $\pm 0.4\,\mathrm{kV}$ uncertainty; even more problematic, the acquisition at the divider with 8 bit oscilloscopes yields a least significant bit of $\sim 0.7\,\mathrm{kV}$. The beam current is measured over a shunt resistor and is known with $0.4\,\mathrm{A}$ accuracy, including the uncertainties mentioned.
A point often neglected is the magnet parameters, where especially the main coil currents have a strong influence. For the W7-X gyrotron magnets, these currents lie in the approximate range 80–$85\,\mathrm{A}$ and can be measured with an accuracy better than $0.1\,\mathrm{A}$. Even though this is a better relative accuracy than the beam current measurement, here a more severe uncertainty is introduced; Ω_r depends directly on the magnetic field, and therefore on the coil current. For a main coil current $I_\mathrm{m} = 83.4\,\mathrm{A}$ and $\gamma = 1.13$, the cavity electron-cyclotron resonance frequency is $\Omega_\mathrm{r}/2\pi = 137.865\,\mathrm{GHz}$. With a pessimistic main coil current uncertainty of $\pm 0.1\,\mathrm{A}$, Ω_r is only known to lie in the range 137.700–$138.030\,\mathrm{GHz}$.
A basic problem in operating point description is therefore that the values are specified more accurately than they can be measured.

3.1.2 Frequency and spectrum measurement

General

Frequency is a physical quantity which can be measured with extreme accuracy; possible techniques for this were already discussed in Section 1.3. Note that while FFT-based approaches and mixer-enhanced spectrum analyzers easily allow frequency determination down into the sub-kHz range, this is only meaningful if the signal to be measured has corresponding stability.
A second important property of the frequency measurement at high power gyrotrons is, that it does not easily allow a judgment of the absolute RF power based on the receiver power. In the broadest sense, the reasons for this lie in the transmission and conversion techniques for the involved measurement signals.

The most relevant factors to be considered are as follows:

- **Pick-up**
  The RF signal at the gyrotron is picked up by a small conical horn antenna or an open waveguide. Because of the high RF power involved, the pick-up is normally located at the far edge of the main RF beam or a relief window of the tube. This causes the signal to be picked from the stray radiation in the gyrotron, which can be assumed to be of roughly homogeneous intensity and unpolarized. As the absolute stray radiation level can fluctuate significantly depending on the operating point, this introduces an important uncertainty.

- **Long transmission line**
  Since the measurement equipment is typically located several meters away from the gyrotron itself, electrically long transmission lines are required. To lower the attenuation, these are normally overmoded; the ensuing dispersion and geometrical effects easily cause a transmission response variation of the order of 10 dB, depending on the implementation.

- **Frequency conversion losses**
  Since the harmonic mixers employed by heterodyne systems are wide-band models, not much optimization for flat conversion response is possible. This causes also the mixers to significantly contribute to frequency-dependent receiver power variations.

Stray radiation and receiver protection

The nominal cavity mode is mostly coupled out of the gyrotron, forming the main beam. For a typical modern gyrotron, more than 97% of the generated main mode power is available as directed beam power (compare e.g. [DAA^{+}02]). For a megawatt gyrotron this means at most a stray radiation contribution of $\sim$30 kW from the main mode. If another cavity mode is operating, its relative stray radiation contribution can be significantly larger, as it is scattered in the launcher to a certain

extent. On the other hand, its total generated power is typically 3–6 dB below the nominal mode's. Parasite powers are hard to estimate, but usually they can be expected to be below 10 kW. It is important to note that since they are not matched to the launcher, they are practically converted into stray radiation completely.
A consequence of these connections is that the ratio between generated and received power at a millimeter-wave probe differs for signals depending on their source, preventing meaningful comparison of measured values. The nominal mode will have a low ratio, while parasites will have a very high one. However, consider that the nominal mode is the strongest oscillation, a non-nominal cavity mode is normally weaker, and the parasites generate the smallest amount of power; this is a highly welcome effect, as it tends to compress the dynamic range of the incident signals.
During experiments, significant power changes can occur, for example through mode switches. Another problem is that a strong signal outside of the receiver band might be unnoticed, resulting in harmful sensitivity settings. An amplitude level control system was developed, which can protect the front-end of a wide band measurement systems, and is documented in [Nus11]. It covers the input frequency range of 110–170 GHz with a dynamic range of 30 dB and a speed of 14 kHz, using an electronic attenuator and a broadband diode detector.

3.1.3 Power measurement

In this section, the means for power measurement in the different pulse domains (compare Section 2.4.3) are reviewed. Since in this work simulated and measured power are to be compared, the goal is the introduction of the necessary translations between both domains.

General classification

In short pulse operation, a ballistic calorimeter mounted directly at the microwave window allows relatively precise measurements in the range $t_\mathrm{p} = 1$–$10\,\mathrm{ms}$. A fraction P_ref of the total power P_tot exiting the main window is reflected by the calorimeter; the power reflection coefficient $|\underline{r}|^2$ defining this fraction is specified in [Sti88] to lie within the range of 0.03–0.06. Apart from this, the power absorbed by the calorimeter is identical to the total power leaving the main window cross section.

For long pulses, a water cooled CW load must be employed, which takes many seconds to stabilize thermally. This severely limits the possibilities for power measurements in the intermediate range $10\,\mathrm{ms}$ to $1\,\mathrm{min}$. It should also be noted that because of its size and the cooling complexity, the LP load is connected to the gyrotron by means of a quasi-optical transmission line consisting of metallic mirrors. An overview on the KIT LP setup can be found in [SEG$^+$11].

While the SP load absorbs all power available at the output window except the reflected fraction, the transmission line between the window and the long pulse load acts as a filter: Modes other than the fundamental Gaussian will be diffracted and therefore converted to stray radiation in the load box. Since some of the mirrors incorporate fractional-λ-depth gratings for polarization conversion, transmission is also reduced for signals deviating from the design frequency. The LP load therefore mainly detects the part of the generated power which leaves the gyrotron as a linearly polarized $\mathrm{TEM}_{0,0}$ beam close to $140\,\mathrm{GHz}$.

Since the simulation tools introduced in Section 3.2.2 predict the RF power at the end of the cavity P_{diff}, measured power values must be corrected for a qualified comparison.

LP power balance

During LP operation, enough energy is deposited in the individual cooling circuits of the setup to enable more detailed power balancing than in SP mode. This allows the cooling circuits of the gyrotron itself, from collector over the RF geometry and internal mirrors down to the microwave window, also the external mirrors, and other secondary beam dumps to be evaluated for a total power balance. If done over the whole pulse's time, the total generated power can be routinely

tracked from the individual cooling circuits with an accuracy in the range of a few percent of the DC input power $P_{\mathrm{dc}} = V_{\mathrm{cath}} I_{\mathrm{b}}$ [Thu03]. An even better accuracy of about 1% can be reached over the direct use of ΔT-values from the individual components in shorter "calm" time segments of the pulse.
The method is slightly complicated by the fact that a number of cooling circuits are implemented in combination, e.g. the absorbed power in the launcher and the first internal mirror cannot be separated. For the method described here, only the combination of cavity and shaft cooling are problematic; we refer to this power fraction as P_{cav}. The rest of the internal RF cooling circuits, namely the mirrors and the window, are summarized here as P_{int}, excluding the collector circuit since its absorbed power is vastly dominated by the terminating electron beam. Combining all external circuits except the main RF load in P_{ext}, the formulation of a summary equation for the total generated RF power P_{gen} is possible:

$$P_{\mathrm{gen}} = P_{\mathrm{dc}} - P_{\mathrm{coll}} \approx P_{\mathrm{cav}} + P_{\mathrm{int}} + P_{\mathrm{ext}} + P_{\mathrm{load}} \quad . \tag{3.2}$$

The power available at the cavity output is then estimated as

$$P_{\mathrm{diff}} \approx P_{\mathrm{gen}} - P_{\mathrm{cav}} = P_{\mathrm{int}} + P_{\mathrm{ext}} + P_{\mathrm{load}} \quad , \tag{3.3}$$

which can be directly used for the assessment of LP measurements with simulation results.

SP power correction

In SP operation, the mentioned multitude of cooling-circuit sensors is not available. The only direct power indicator is the ballistic calorimeter, which yields a measured average power P_{meas} for a complete pulse. Three important components of this measured value should be considered when relating to simulations:

1. A systematic correction for the RF losses occurring between the cavity exit and the atmospheric side of the vacuum window is necessary.

2. The reflectivity of the calorimeter introduces uncertainty and a systematic offset.
3. From pulse-to-pulse reproducibility experiences, an absolute uncertainty $\pm\Delta P_{\text{rep}}$ of less than $\pm 40\,\text{kW}$ can be expected.

The internal RF losses can be described as a factor s_{int} connecting the power P_{out} available at the window cross section with the power P_{diff} leaving the cavity. As this factor should be constant for a given mode, geometry and reasonable frequency range, it can be found by investigating the LP power balance. Using data from several long pulse shots of four different W7-X gyrotrons, this factor could be determined to be in the range

$$s_{\text{int}} = \frac{P_{\text{diff}}}{P_{\text{out}}} = \frac{P_{\text{ext}} + P_{\text{load}}}{P_{\text{int}}} = 1.031 \pm 0.004 \quad . \tag{3.4}$$

As the reflectivity of the calorimeter is specified to lie between the values $|\underline{r}_{\text{cal,min}}|^2 = 0.03$ and $|\underline{r}_{\text{cal,max}}|^2 = 0.06$, an additional offset factor of

$$s_{\text{ref}} = \frac{1}{1 - \overline{|\underline{r}|^2}} \approx 1.047 \pm 0.016 \tag{3.5}$$

is to be expected.[1]
Using these estimations, a power correction factor of

$$s_{\text{corr}} = s_{\text{int}} \cdot s_{\text{ref}} \approx 1.080 \pm 0.021 \tag{3.6}$$

can be derived, which relates the measured power to an estimated diffracted power as

$$P_{\text{diff}} \approx \overline{P}_{\text{corr}} = \overline{s}_{\text{corr}} P_{\text{meas}} \quad . \tag{3.7}$$

Applying the conservative uncertainty ranges stated above, plus the empirical reproducibility limitation ΔP_{rep}, a measurement confidence

[1] In principle the induced uncertainty could be larger, as a standing wave can form between a reflective power load and a reflective source. Since the calorimeter will induce significant diffraction in the reflected wave, and the window has a much lower reflectivity than $-30\,\text{dB}$, this further complication was ignored here.

corridor as

$$P^{+}_{\mathrm{corr}} = \max(s_{\mathrm{int}}) \cdot \max(s_{\mathrm{ref}}) \cdot P_{\mathrm{meas}} + \Delta P_{\mathrm{rep}} \quad (3.8)$$

$$P^{-}_{\mathrm{corr}} = \min(s_{\mathrm{int}}) \cdot \min(s_{\mathrm{ref}}) \cdot P_{\mathrm{meas}} - \Delta P_{\mathrm{rep}} \quad (3.9)$$

is suggested. If $P^{-}_{\mathrm{corr}} \leq P_{\mathrm{sim}} \leq P^{+}_{\mathrm{corr}}$ holds, the simulation can be considered to match with the underlying experimental result.

3.1.4 Example measurement

In the following, a wide experimental sweep over the cathode voltage of the W7-X gyrotron SN6 in SP operation is presented. This sweep will serve as a reference scenario for the various interaction analysis methods described subsequently.
Figure 3.1 contains the relevant data of the sweep, which was done during parameter studies in short-pulse operation as a first step of the factory acceptance test at the KIT high power gyrotron test stand.
In the top graph, the variation of α and I_{b} versus V_{cath} and the effective beam voltage in the cavity $V_{\mathrm{b}} = W_{\mathrm{k}}/e$ is visible.[2] While the changes in V_{b} and the pitch factor are the main sweep variables, the beam current increase is caused by the Schottky effect. The bottom graph displays the measured power and frequency of each operating point covered in the sweep from 70 to 90 kV. A relatively linear RF power increase with V_{cath} is observed, ending with a sharp drop as the azimuthal neighbor mode takes over at $V_{\mathrm{cath}} \approx 88\,\mathrm{kV}$, which is expected behavior. In addition, a power corridor is plotted, which represents an estimation of the power available at the cavity exit, as calculated with the relationships laid out in Section 3.1.3.
With rising V_{cath}, the RF frequency drops as the relativistic cyclotron frequency is reduced by the increase in γ. For cathode voltages above 82.5 kV, weak parasitic oscillations in the 130 GHz range were noted during the experiment, which however did not seem to influence the behavior significantly. The sweep presented is not a high-power operation sweep; it was rather selected for as "mild" behavior as possible, as it serves as the basis for a benchmark.

[2]The values of α and V_{b} were simulated with ESRAY [IB99].

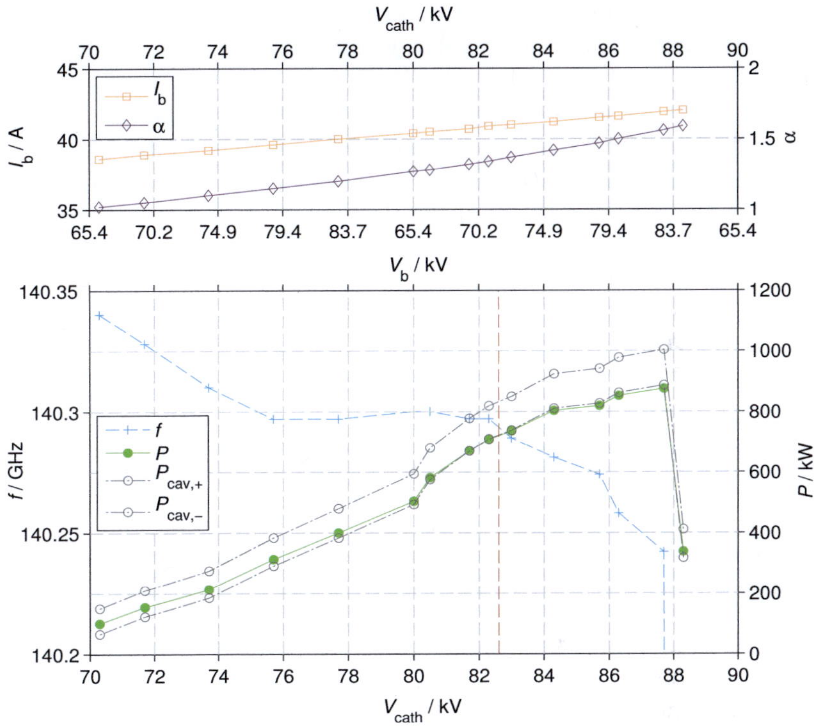

Figure 3.1: Sweep over cathode voltage during gyrotron optimization
Top: I_{b} and pitch factor α versus V_{cath} and V_{b}
Bottom: measured frequencies, powers and corresponding cavity power corridor for simulation results

3.2 Simulation techniques

3.2.1 Beam parameters

The magnetic flux density and the beam parameters cannot be measured during an experiment, and must be simulated. At KIT, this task is performed by the in-house codes MAGGEN / ESRAY [IB99] and ARI-

ADNE [PV04]. A verification of these beam parameters is only given in subsequent interaction simulations and their agreement with the experiment, which is a fragile and indirect way of validation.

B_z is calculated from the magnet properties, that is the geometry and the coil currents; for W7-X gyrotrons the relevant external parameters are I_m for the main solenoids and I_{tr} which modifies the gun field. However, systematic deviations were noticed in a comparison of on-axis B_z profiles, measured using several Hall-probes, with simulated results. A fitting approach which allowed shifts of the magnet coils could significantly improve the match by introducing offsets in the 0.1–$3\,\mathrm{mm}$ range [Ill11]. The explanation for this lies very likely in thermal strain effects during the transition of the dewar from room temperature to $4\,\mathrm{K}$, so the fit corrects the "warm" design geometry with respect to the true "cold" geometry which determines the measurements.

In a nominally operating gyrotron the beam parameters as introduced in Section 2.1 are dominated by the gun geometry and the surface properties. Recent modeling advancements at KIT have now included a more physical treatment of the emitter surface roughness, but are still under refinement [ZIPJ13]. The basic approach is the introduction of macro-bump structures distributed on the emitter surface. These are constituted by different elementary geometrical forms such as concave and convex semi-spheres, semi-cylinders and pyramids, characterized by a size parameter h. In the Figure 3.2 the related beam parameter variation for a feature size range of $0 \leq h/\mu\mathrm{m} \leq 15$ is plotted for two sample W7-X gyrotron operating points. Both points are set at $I_b = 40\,\mathrm{A}$, one slightly above nominal at $V_{cath} = 83\,\mathrm{kV}$ and the other one at the edge of the stability range of the $\mathrm{TE}_{28,8}^{-}$ mode at $87.7\,\mathrm{kV}$.

As expected, the parameter spread increases with the emitter roughness, and interestingly the dependence $\partial(\delta\alpha)/\partial h$ is strongest at the onset, $h = 0^+$. Initially identical, the simulated spreads separate for the two operating points with increasing h. Apart from the spreads, the mean parameter values also rise with the feature size. Not shown in the figure is the form of the statistical distributions, which for values $h > 10\,\mu\mathrm{m}$ noticeably deviates from a Gaussian distribution.

In the frame of the simulation study presented in Section 3.2.2, different

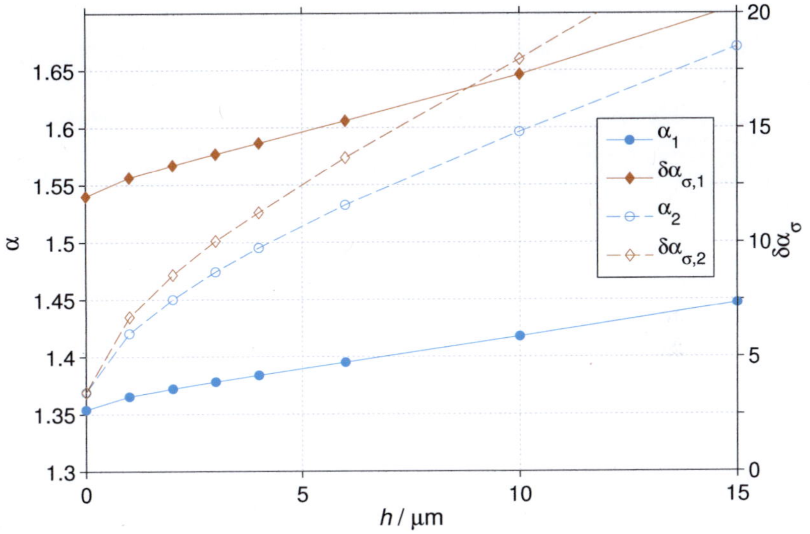

Figure 3.2: Example of the beam parameters ($\alpha = \overline{\alpha}$) at two V_{cath} operating points ① and ② over emitter roughness

emitter roughnesses were also investigated. This does not replace a systematic study of the roughness effect on interaction simulations, and is also beyond the frame of this work. The results indicate that a conservative value of $h \gtrsim 2.5\,\mu\mathrm{m}$ can be assumed, which agrees well with preliminary optical inspection [Zha14].

3.2.2 Gyrotron interaction analysis

Linear interaction calculation

The primitive resonance condition (2.21) presented in Section 2.2 can be used to calculate the gyro-oscillation frequency of a specific mode for a given waveguide radius r_{w} and magnetic flux density B_{z}. Its limitations (see Section 2.2.3) make the result a relatively coarse estimate. There exist also approaches which attempt to bring these analytical

results closer to the actual values, mainly by taking into account the effect of the electron beam and the modifications it causes in the mode dispersion relations. These techniques can be termed *coupled dispersion relations* (CDRs), to be distinguished from the *uncoupled dispersion relation* (UDR) as introduced in Section 2.2. While in the UDR, the point of interaction is defined by the intersection of a beam line with a dispersion hyperbola, in CDR the hyperbola is replaced by a multi-branch structure. Additionally a growth rate term over $k_{\|}$ is achieved, which defines probable points of interaction with its local maxima. However these linear calculations are not able to give any indication of the expected RF power.
CDRs for gyrotrons were developed in [Edg80, CA81], but rather from a design point of view and without verification or qualification of the results. A more extensive treatment of the problem is given in [ZSS07], where the presence of positively charged plasma is considered in addition.[3] Even though the nominal parameters of the W7-X gyrotrons are used in this publication, unfortunately only qualitative agreement with other tools and experiments is stated, without further quantification.

Self-consistent slow time interaction simulation

The main application goal of the codes of this category is to be the central tool for gyrotron design and to perform the necessary parameter studies, requiring reasonable computation times.
Such codes have two very basic techniques in common:

- Basis for the fields modeling is the expansion into waveguide or waveguide resonator eigenmodes. As a consequence, before the simulation start a mode selection must be made, based on linear theory or beam-wave coupling coefficient considerations. If relevant modes are not selected, the simulation cannot reproduce physical reality. Also, in most codes mode conversion is not modeled.

[3] Note that this was motivated by the ionization-based electrostatic neutralization effect, which is described in Section 2.4.3 and Chapter 7 in more detail.

- Instead of evaluating directly the kinetic effect which the electromagnetic fields have on the beam electrons, the trajectories of the electrons are chosen a priori by fixing the electron beam guiding center, i.e. the beam radius. This reflects the focusing by the external magnet, which to a very good approximation forces the electrons to encircle the magnetic field lines in their helical movement.

The key technique to fast simulation however is the "slow-time" formalism, which allows a base-band treatment of the respective mode's eigenfields by interpreting the RF oscillation frequency as a carrier frequency. This allows a fairly relaxed time discretization with respect to the electron cyclotron period, for modern scenarios e.g. in the range $\delta t_{\mathrm{sim}} = 10\text{–}100\,\mathrm{ps}$ for millimeter-wave gyrotrons.
Prominent examples of such so-called time-dependent multi-mode self-consistent beam-wave interaction simulation codes used at KIT are SELFT [Ker96] from the code package CAVITY, and the more recent EVRIDIKI [APIV12] in the code package EURIDICE. Further similar tools are the single-mode code TWANG [ATA+11], MAGY [BAL+98] and the TUHH[4] code [JGS99, Jel00], which includes more sophisticated physical modeling, sacrificing computational speed.
An important additional simplification used by all codes except the latter one is an extension of the slow-time formalism, which assumes the RF envelope to change much more slowly than the electron transit time through the considered geometry. Thus the code can propagate the electrons along their complete trajectory through the simulation region before performing an update of the RF fields by integration. This also gave raise to the term "trajectory codes".[5]
Despite high computation speed and design-friendly use, the codes are subject to important limitations. These are mainly consequences of the techniques mentioned above, and are summarized in a comprehensible

[4]Technical University Hamburg-Harburg

[5]By this extension, implicitly a further time constant is introduced into the simulation, which has more complicated properties than mere sampling: For e.g. a simulation region with an axial extent $L = 100\,\mathrm{mm}$ and $\beta_{||} = 0.3$, the cavity transit time is $\delta t_{\mathrm{el}} \approx 1\,\mathrm{ns}$ while the RF fields are updated with the periodicity $\delta t_{\mathrm{sim}} \ll \delta t_{\mathrm{el}}$.

way in [APIV12]. For example rotational symmetry of the geometry and co-axiality with the electron beam are fundamental properties, but adaptations for investigations of mild beam eccentricity have been made [APIT13, DN13]. Many restrictions apply to axial dependences: Due to the classical TE/TM mode treatment without mode conversion, only smooth geometry tapers $\lesssim 5°$ and high wall conductivity are allowed in most codes.
The axial variation of the external static magnetic field and its impact on the beam parameters are another important issue. Typically not all consequences of $B_z(z) \neq \text{const}$ are considered; e.g. $\Omega_r(z)$ and $r_b(z)$ can be included at relatively low computational cost, while the impact on the momenta and especially $\beta_{\|}$ is neglected. In [Cho14], a modification of SELFT addressing this matter is presented, and a similar extension has been implemented in EURIDICE.
For the oscillation frequencies involved, the general limitations are as follows:

- The electron beam is represented by a discrete number of beamlets composed of macro-electrons. Only wave oscillations with an instantaneous frequency f_{RF} close to f_{cyc} within a few percent (i.e. gyrotron-type interactions) are modeled reliably.

- The "slow-time" approach is basically a baseband transition, relying on an internally defined numerical carrier frequency f_{carr}. Since the baseband RF envelope fluctuates with the difference frequency $|f_{RF} - f_{carr}|$, the more an oscillation's instantaneous frequency deviates from this carrier, the less the slow-time approximation holds.

An important difference between the SELFT code family and EVRIDIKI is the modeling of the electron phase angles and the beam parameter spread. A comprehensive overview would exceed the scope of this work; however, it should be stated that in SELFT, the batches of electrons entering the simulation domain from upstream are identical. Inside the batch, the beam parameters can be statistically distributed, but successive batches cannot be distinguished. This requires careful checks to prevent unwanted numerical coupling between the modes over this "pseudo-statistical" electron beam. EVRIDIKI, on the other hand, was built for truly random treatment of the entering macro-electrons, with the advantage of reduced susceptibility to pathological numerical effects, and a disadvantage in computing performance.

Particle-in-cell (PIC) simulations

Various different concepts are assigned this label, which have in common that the "slow time" assumption is mitigated or entirely removed. This significantly increases the required computational resources as compared with slow time codes, which is acceptable since the main goal of these codes is the investigation of physical effects and the validation of the fast design codes and their assumptions.
"Full-PIC codes" do not contain a mode-based field description, but operate on the terms of the Maxwell equations more or less directly, typically applying finite-difference time-domain (FDTD) schemes. This entails very high demands on time and frequency discretization, which are one reason for the comparably low speed. Until recently, full-PIC codes were out of the question for gyrotron simulation because of the high complexity and size of the necessary scenarios. However in [SNR+12], a successful interaction simulation of a $TE_{0,3}$ mode gyrotron with the code HALO was presented in great detail and comprehensiveness. The same code is applied in [NSF+12] to a current state-of-the-art problem, the investigation of the mode conversion in a $TE_{34,19}$ mode 170 GHz launcher, underlining the flexibility of a full-PIC approach. Despite the obvious advantages of full-PIC calculations over more simplified approaches, they cannot simply be regarded as the solution for all simulation tasks, as they bring along their own inherent problems. These typically concern the proper formulation of boundary conditions, violation of energy conservation, convergence issues and the nontrivial a-posteriori task of meaningful mode analysis.
A number of codes called "quasi-PIC" also exist, one sub-category of which consists of the "two-and-a-half-dimension" PIC codes. These do still include azimuthal symmetry, and can be placed in between the

design codes and the full-PIC codes. Examples can be found e.g. in the form of a 32 GHz-gyrotron interaction simulation in [BKC99], or in [IB99], where also electrostatic electron beam plasma instabilities in a gyrotron are investigated with the code TH2PIC. This code was also used for a study of few operating points in a $\text{TE}_{0,3}$ gyrotron, which served as a benchmark for the design code developments in [Ker96, JGS99].
The other quasi-PIC sub-category stems from advancements of slow-time design codes, where among other things changes of the electron propagation technique and intermediate field updates during the transit allow the introduction of PIC-style features. Such an extension is available in EVRIDIKI. Of course basic properties such as the idealized eigenmode treatment or the confinement of electron trajectory centers to the magnetic field lines remain. Advantages lie in the possibility to reduce unwanted effects such as unphysical mode excitation and coupling, which arise if the slow-time assumptions in the design codes are violated.

3.3 Benchmarking of experiment and simulation

In the following, the different tools for gyrotron frequency and power estimation are benchmarked against the measured data in the V_{cath} sweep of Section 3.1.
The primary quantities for a benchmark at a given operating point are the main oscillation frequency f and the RF output power P. Because of the many factors which can lead to disagreement between simulated and measured data, all benchmarks in this work were performed as sweeps over multiple operating points, which allows the investigation of parameter tendencies in addition to the absolute values. While, for example, the operating mode $\text{TE}_{m,n}$ cannot be directly determined in the experiment, the position of cavity mode transitions in parameter sweeps are marked by discrete changes of output frequency and power. As mentioned, weak parasitic signals were detected for $V_{\text{b}} \gtrsim 77\,\text{kV}$, which are not of interest in this study. In all graphs, the onset beam voltage is marked with a vertical dashed line for reference.

The magnetic field and the beam parameters were calculated separately for each simulated operating point, following the reasoning in Section 3.2.1.

3.3.1 Frequency calculation results

For each operating point, the frequencies predicted by the CDR and a UDR implementation were calculated (compare Section 3.2.2), with electron beam parameters gathered from the KIT in-house code ESRAY. To address the span of local beam parameter modification by the dynamic interaction mentioned earlier, the parameters for the "used beam" in the cavity were also estimated. In an empirical approach, $P_{\mathrm{RF}}/2$ was chosen, in order to emulate an operating point "halfway" between the fresh and the spent beam (see also Section 2.2.3).
The results are displayed in Figure 3.3, along with the relativistic tuning of the electron cyclotron frequency through γ on the top axis. Obviously the measured RF follows the tuning of Ω_{r}, but with nothing close to the same slope. The direct application of the UDR is disillusioning, as its discrepancy with the measured values is hundreds of MHz, up to more than 2 GHz for the peak power operating point. For the CDR, the difference is surprisingly larger and not smaller, indicating that it is not the space charge of the beam which causes the divergence between measured results and linear theory. Both techniques do not even qualitatively agree with the true frequency variation, since they predict a roughly hyperbolic frequency development with a linear increase of V_{cath}.
At most V_{cath} points the calculated frequencies are too high, so no improvement from the excitation band considerations (see Section 2.2.3) can be expected, as this would only further increase the frequencies by some hundred MHz. The values obtained with the estimated "spent beam" parameters interestingly yield smaller discrepancies. For the UDR as well as the CDR, the unreasonable parabolic inclination is mitigated, and closer agreement with the measurement is reached. The applicability of this empirical correction is limited by the need for power values, which makes it an a-posteriori method.

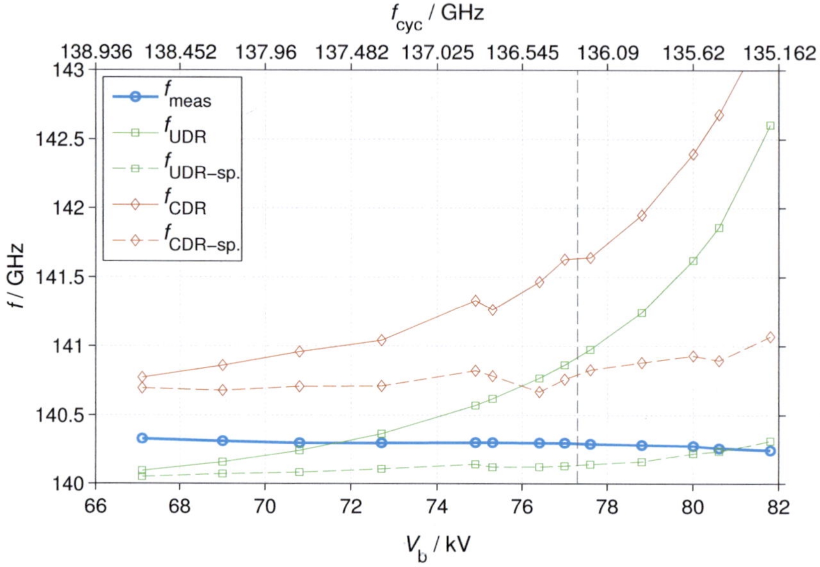

Figure 3.3: Measured frequency from Fig. 3.1 and analytical solutions; the "sp" suffix denotes the spent beam modification.

3.3.2 Self-consistent simulation codes and variants

Of the simulation tools described in sections 3.2.2, five different variants are benchmarked:

- SELFT, the design code at KIT from the CAVITY package [Ker96]
- SELFT-BZ, an extended SELFT with better $B_z(z)$-treatment [Cho14]
- EVRIDIKI, the parallelized design/analysis code from the EURIDICE package [APIV12]
- EVR-BZ, an extension of EVRIDIKI for $B_z(z)$-treatment
- EVR-PIC, the quasi-PIC variant of EVRIDIKI (including also $B_z(z)$ treatment)

In Table 3.1, a list of the tools with sample performance times for the benchmark simulation setup is supplied.[6] While the non-parallelized SELFT code has the best per-thread performance, the parallelized EVRIDIKI codes give the possibility of considerable speed gain by cluster computing. Because of this, for SELFT and SELFT-BZ an in-house workstation was employed, whereas the EVRIDIKI-based calculations were mostly made on the HELIOS-cluster[7].

The most ambitious simulation code in this comparison is the quasi-PIC approach EVR-PIC. For the studies presented, the start-up simulations at each operating point were only carried out with the non-PIC codes. EVR-PIC was invoked only in continuation runs, i.e. simulations based on the final values from another simulation, in this case EVR-BZ. Interestingly, the speed penalty for the enhanced consideration of z-varying magnetic flux density is considerably smaller in the

[6]The performance times are only a general illustration; they heavily depend on the number of modes, macro-electrons and the spread modeling settings. Naturally the performance of the EVRIDIKI codes does not scale linearly with the number of threads employed due to multi-threading and cluster inter-node communication.

[7]A system composed of 4410 computation nodes with a peak performance of 1.52 Peta-Flops and a usable memory of 256 TB. Each node contains 16 cores and 58 GB of RAM. It is operated by the International Fusion Energy Research Centre (IFERC) in Rokkasho, Japan [http://www.iferc.org/csc].

Tool	$\delta t/\text{ps}$	t_{sim}/ns	t_{real}/h	n_{th}	$\frac{t_{\text{real}}}{t_{\text{sim}}} / \frac{\text{min}}{\text{ns}}$	$x / \frac{\text{min}}{\text{ns}}$
SELFT	10	2500	21	1	0.50	0.50
SELFT-BZ	10	2500	155	1	3.73	3.73
EVRIDIKI	10	2500	6.6	64	0.16	10.12
EVR-BZ	10	2500	7.9	64	0.19	12.19
EVR-PIC	1	300	5.7	80	1.14	90.90

Table 3.1: Simulation codes with sample performance data: timestep δt and length t_{sim} of simulation, real time computation duration t_{real}, number of used parallel threads n_{th}, total speed $\frac{t_{\text{real}}}{t_{\text{sim}}}$ and thread-weighted speed $x = \frac{t_{\text{real}}}{t_{\text{sim}}} \cdot n_{\text{th}}$

EVRIDIKI family than in the SELFT group. This can be seen from the ratios between simulation time and its real time computation duration, $t_{\mathrm{real}}/t_{\mathrm{sim}}$, in the table: While SELFT-BZ takes 7.5 times the duration of SELFT, the factor between EVR-BZ and EVRIDIKI is only 1.2.
In the following, first the simulation setup is presented, along with two sample start-ups which were performed with two different simulation codes. Then, a comprehensive reproduction of the benchmark measurement from Section 3.1.4 with the different simulation tools is performed and discussed thoroughly.
It should be noted that uncertainties in the main magnet coil current I_{m} can be used to justify slight tuning of the magnetic field in the simulation, in order to achieve better agreement with measured values (compare Section 3.1.1). This degree of freedom was not used in any of the presented simulations deliberately, as it could mask important deviations. For similar reasons, the numerical carrier frequency was kept constant.

3.3.3 Scenario

As simulation domain, the complete cavity including the downtaper and full nonlinear uptaper is chosen. In order to reproduce the experiment as well as possible, each operating point was simulated with a parameter start-up ramp, emulating the experimental situation. Applying this "start-up" in simulation allows the sequential excitation of transient modes, which can be fundamentally influential to the final stationary result through hysteresis effects.[8]
The starting point of the 2 µs duration ramps is $V_{\mathrm{cath}} = 25\,\mathrm{kV}$, with the beam current before flat-top following the Schottky-effect. As the flat-top end values lie in the range 67–82 kV, slew rates of 20–30 kV/µs are attained. This is still two orders of magnitude above the typical experimental values; however, considering the electron transit time of a few nanoseconds, both the simulation and the experimental parameter ramp are very gently inclined with respect to the gyrotron interaction time constants.

[8]Compare also Section 2.4.1.

Mode selection

A multi-mode selection was made based on the beam-wave coupling coefficients in the cylindrical cavity section z-range, using the CAVITY component SCNCHI.
Typical criteria for the mode selection are

- z range, $\Delta z = [z_1, z_2]$,
- cut-off frequency range, $\Delta f_{\mathrm{cut}} = [f_{\mathrm{cut},1}, f_{\mathrm{cut},2}]$ and
- minimum relative coupling coefficient $Ge^{\mathrm{r}}_{\mathrm{min}}$.

The latter is obtained by normalizing the individual curves $Ge_{m,n}(z)$ with the coefficient of the design mode at the center of the cylindrical cavity section $z_{\mathrm{cav},0}$, as

$$Ge^{\mathrm{r}}_{m,n}(z) = \frac{Ge_{m,n}(z)}{Ge_{28,8}(z_{\mathrm{cav},0})} \quad . \tag{3.10}$$

For the simulations presented in the following, only modes with $\max\left(Ge^{\mathrm{r}}_{m,n}(\Delta z)\right) \geq 0.6$ were considered, where Δz was limited to the cylindrical cavity section. The cut-off frequency range was set to $125 \leq f_{\mathrm{cut}}/\mathrm{GHz} \leq 155$.
A list of the modes considered relevant by these criteria is provided in Table 3.2; graphical representations of the simulation geometry and the relative coupling coefficients in the cavity can be found in Figure 3.4.

$+18,10$	$(+24,8)$	$-26,7$	$-27,8$	$\mathbf{-28,8}$	$-29,8$	$-30,9$
$+19,10$	$+24,9$	$(+26,8)$	$+27,8$	$+28,8$	$-29,9$	$-31,8$
$(+23,8)$	$(+25,8)$	$+26,9$	$+27,9$	$+28,9$	$+29,9$	
$-23,10$	$+25,9$	$(-27,7)$	$[-28,7]$	$-29,7$	$-30,8$	

Table 3.2: TE mode list for simulations; modes in ()-brackets were not considered in the final simulation, the $\mathrm{TE}^{-}_{28,7}$ mode is discussed in the last paragraph of Section 3.3.7.

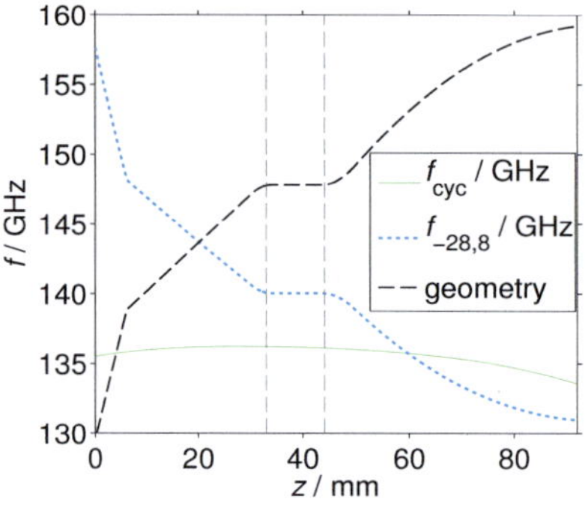

(a) Cyclotron frequency, main mode cut-off frequency $f_{-28,8}(z)$, and geometry

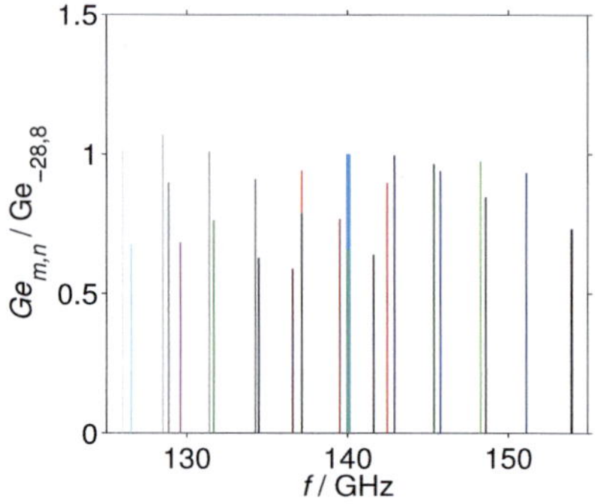

(b) Relative coupling coefficients vs. cut-off frequency in cylindrical section

Figure 3.4: Overview on the benchmark simulation scenario

Pathological modes

The number of selected modes had to be reduced, as multiple modes of pathological character appeared during the start-up phase in various simulations. These were present as strong forward and/or backward waves far from cutoff, hence penetrating the whole simulated geometry, operating far from Ω_r.

As an example, the $\mathrm{TE}^+_{24,8}$ mode can dominate the mode spectrum for $400 < t_\mathrm{s}/\mathrm{ns} < 900$, oscillating between 145 and 150 GHz. Its cut-off frequency at the cylindrical section is 128.7 GHz, and at the minimum radius at the geometry entrance, $f_\mathrm{cut} \approx 135\,\mathrm{GHz}$. With $f - f_\mathrm{cut} > 10\,\mathrm{GHz}$, this resembles a gyro-TWT interaction. The physical relevance of these oscillations is disputable as they violate the code limitations listed in Section 3.2.2, but also from a practical point of view they impose a problem: Typically these modes dominate the start-up sequence for a relevant length of time, and they were found to be able to grossly influence the final stationary state after the end of the start-up.

At the operating point presented, the pathological behavior was mostly, but not exclusively, observed in $\mathrm{TE}^+_{m,8}$ modes. They were removed from the list step by step: As soon as a problematic mode emerged, a new start-up without the pathological mode was launched. In Table 3.2 the affected modes are set in brackets.[9]

Despite both code families relying on the same technique, the SELFT-based codes were found to be more susceptible to this effect than the EVRIDIKI-based ones. The problem appeared in the same modes, however, the impact was much smaller in EVRIDIKI. An explanation for this could be the different implementations of the phase discretization and parameter spread modeling, as mentioned in Section 3.2.2. EVR-PIC did not show this problematic mode behavior, even when running as a continuation of an affected EVR-BZ start-up. This supports the doubts about the physical reality of these modes.

3.3.4 Sample start-up with SELFT

In Figure 3.5, the mode time evolution of the simulation with SELFT at the operating point $V_\mathrm{b} = 77.0\,\mathrm{kV}$ (with $I_\mathrm{b} = 41.7\,\mathrm{A}$, $\alpha = 1.34$ and $\delta\alpha_\sigma = 8\%$) from the measurement in Section 3.1 is shown. While all modes listed in Table 3.2 without brackets were considered in the simulation, the plot legend only features the modes whose power exceeds 10 W at least at one point of time.

In the beginning, all modes form a uniform noise pattern which is then interrupted by successive starts and declines of various modes. Towards the end of the parameter ramp, the nominal W7-X gyrotron cavity mode $\mathrm{TE}^-_{28,8}$ starts and dominates the spectrum, oscillating with 140.276 GHz and $P_\mathrm{diff} = 763\,\mathrm{kW}$. All other mode activity is limited to powers significantly below 1 kW. This agrees reasonably well with the experimentally measured frequency of 140.297 GHz, and very well with the power corridor of 710–820 kW.

The z-dependence of the electric fields and spectra at the indicated point of simulation time $t_\mathrm{s} = 2400\,\mathrm{ns}$ are provided in Appendix D.1. Again the legend was limited to the relevant modes with respect to

[9]The $\mathrm{TE}^-_{28,7}$ mode is the last mode appearing with pathological behavior. As a direct neighbor of the nominal mode it is problematic to ignore it, so its influence was studied separately and is discussed in the next section.

maximum instantaneous power. The field patterns show how the nominal mode is confined in the cavity, and is limited to the left by the cut-off condition. The weak remnant modes seem to be far away from their cut-off frequency as they extend throughout the entire simulation domain, however operating only marginally above the noise level. In the spectrum, again the clear dominance of the cavity mode is established. The background mode spectrum exhibits discrete line patterns, which are of numerical origin. No beat pattern is visible on the cavity mode in the field display, which would signify dynamic after-cavity interaction (see also Section 6.1.3 and [SCG+11]).[10]

[10] It should be noted that as pointed out earlier, the important neighboring mode $\mathrm{TE}^{-}_{28,7}$ was removed from the selection to obtain this result. A similar simulation including this mode exhibited pathological operation, which resulted in a stationary output power of 667 kW with a frequency of 140.332 GHz, deviating stronger from the measurement.

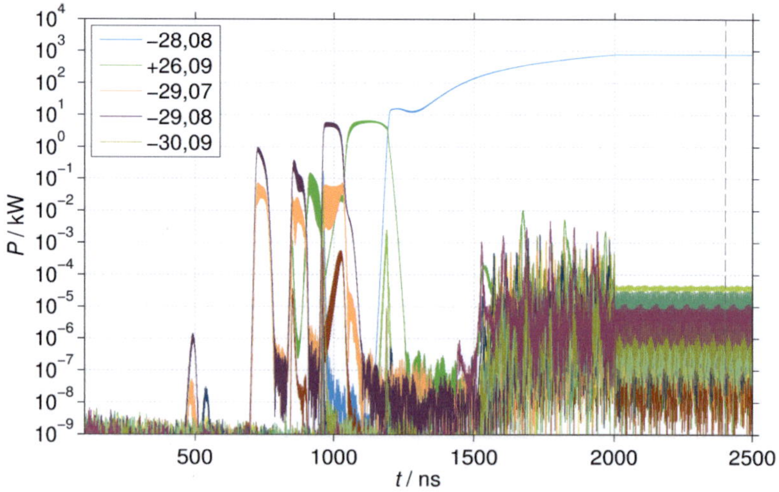

Figure 3.5: Startup simulation of SELFT at $V_\mathrm{b} = 77\,\mathrm{kV}$, $t_\mathrm{ramp} = 2000\,\mathrm{ns}$ with mode selection from Table 3.2 (only relevant modes with $P \geq 10\,\mathrm{W}$ are listed in the legend)

3.3.5 Sample start-up with EVRIDIKI

For the same operating point, the results as simulated with EVRIDIKI is plotted in Figure 3.6. Again stationary operation of the $\mathrm{TE}^{-}_{28,8}$ mode is found, in this case at $f = 140.304\,\mathrm{GHz}$ with $P_{\mathrm{diff}} = 753\,\mathrm{kW}$, which constitutes excellent agreement with the experiment (compare Fig. 3.1). This simulation includes the $\mathrm{TE}^{-}_{28,7}$ mode, which does not rise.
Comparing the mode noise floor of this simulation with the SELFT result, the differences in implementation become apparent: While EVRIDIKI has a relatively high, but rather continuous noise floor, the one in SELFT more inhomogeneous and dependent on the activity of the dominating modes.[11] Similar characteristics can be found in the

[11] Note that the vertical axis scales in the EVRIDIKI and SELFT figures are different, and that the numerical floor phenomena observed in SELFT are below the EVRIDIKI noise floor, due to the different implementation of randomness in the beam parameters.

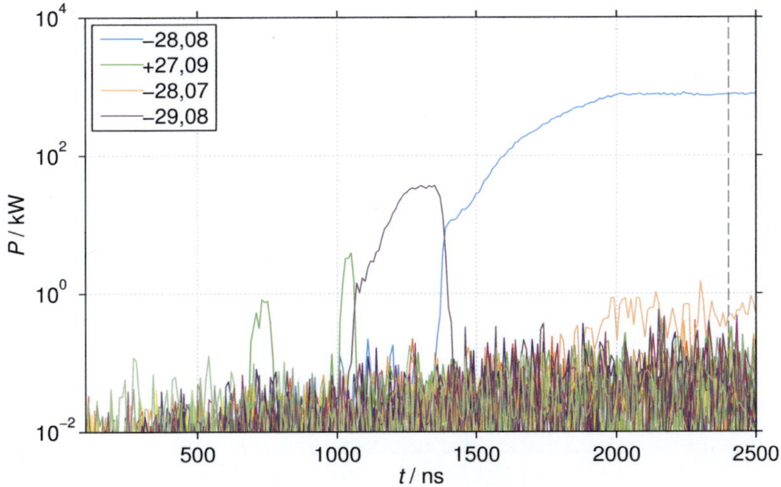

Figure 3.6: Startup simulation of EVRIDIKI with $t_{\mathrm{ramp}} = 2000\,\mathrm{ns}$ at $V_{\mathrm{b}} = 77\,\mathrm{kV}$ with mode selection from Table 3.2 (only modes with $P \geq 1\,\mathrm{kW}$ are listed in the legend)

axial field pattern and the spectrum (see Appendix D.2). Though not explicitly identical, the relevant start-up modes are also similar. In the final steady state, however, a weak $\mathrm{TE}^{+}_{19,10}$ pathological oscillation remains in EVRIDIKI, which appears to be pulled to the main mode's frequency, as visible in the spectrum in Figure D.4.

In the stationary results of the EVR-PIC code, which are given in Appendix D.3 for reference, the delocalized mode far from cut-off is replaced by a very weak $\mathrm{TE}^{+}_{18,10}$ mode oscillation in the cavity.

3.3.6 Benchmark results

All points of the measured sweep from Section 3.1.4 were simulated and investigated in a similar fashion, using the codes listed in Table 3.1. For a meaningful comparison of the measurement and simulation results, the difference values between the two are plotted in Figure 3.7.[12] Instead of the measured power, the central corrected value as introduced in Section 3.1.3 was substituted as $\Delta P = P_{\mathrm{sim}} - \overline{P}_{\mathrm{corr}}$.

The main benchmark consists of the simulation results from SELFT, SELFT-BZ, EVRIDIKI, EVR-BZ and EVR-PIC. As mentioned in the sample results discussion above, the simulations of the latter three codes was performed including the $\mathrm{TE}^{-}_{28,7}$ mode, which however showed no activity at most operating points. The SELFT and SELFT-BZ simulations were performed without this mode, as it caused the results to deviate significantly from the experimental values. This is illustrated by the curve SELFT^{+}, where the mode was considered. For cross-reference, selected operating points were also simulated with EVRIDIKI using the mode list of the SELFT/SELFT-BZ simulation to check the reaction of EVRIDIKI to the removal of the $\mathrm{TE}^{-}_{28,7}$ mode, labeled EVR^{-}. The sweeps SELFT^{+} and EVR^{-} are not considered to be part of the main benchmark, but are discussed separately to give more insight on the relevance of the pathological phenomena.

[12] For all operating points in this sweep, the same magnetic field was used directly as simulated by MAGGEN, without any fine-tuning within the coil current uncertainties.

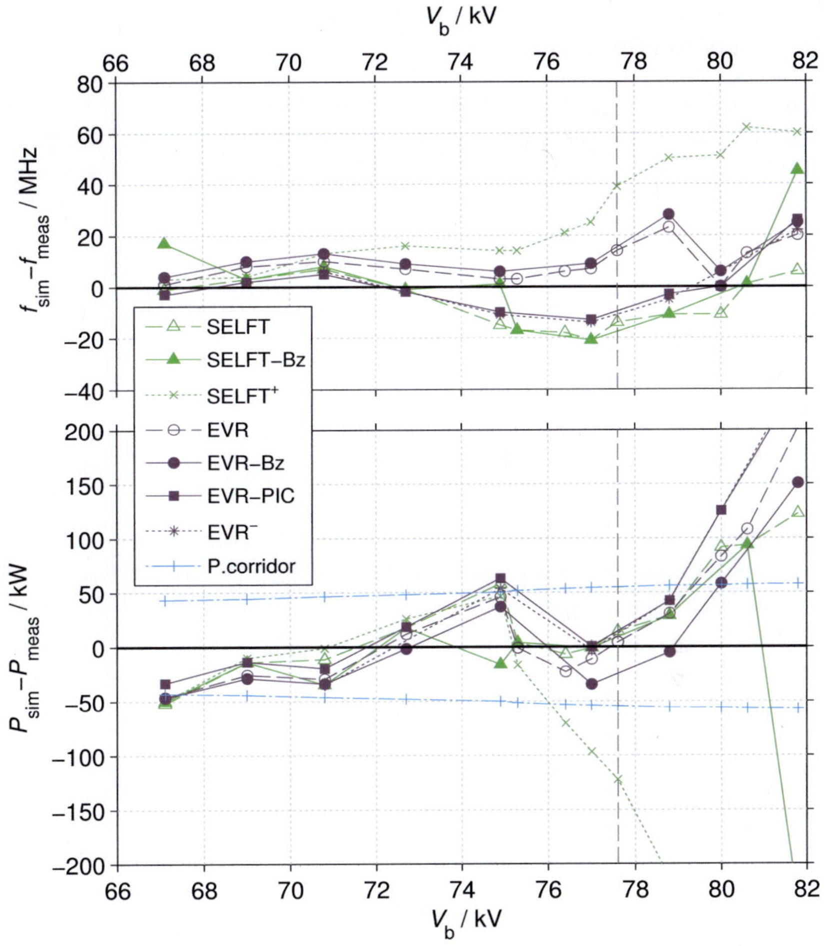

Figure 3.7: Frequency deviation (top) and power difference (bottom) between simulation and experiment;
P.corridor denotes the uncertainty from measurement (see Section 3.1.3; the dashed vertical line marks the onset of weak parasites in the experiment (compare Section 3.1.4)

3.3.7 Discussion

Main benchmark sweep

The frequencies from the experiment are reproduced very well by the simulations, especially in comparison to the results of the linear calculations discussed above in Section 3.3.1. For the predominant part of the sweep, the frequency deviation remained in $|\Delta f| < 20\,\mathrm{MHz}$ with all codes considered. Above beam voltages of 72 kV, a formation of two groups can be observed: While EVRIDIKI and EVR-BZ tend to overestimate the frequency, SELFT and SELFT-BZ do the opposite.[13] In particular, EVR-PIC agrees well with the latter codes.
Considering the power corridor also present in the lower part of the figure, the power is predicted within the expected uncertainty up to $V_\mathrm{b} = 78.8\,\mathrm{kV}$. Above this point, the codes increasingly overestimate the RF power. In Figure 3.1 one can see that the pitch factor approaches and exceeds 1.5 in this range, a situation in which the gyrotron cannot be expected to operate reliably. It is very well possible that here multiple effects come into play which are not covered in the simulation.
In the experiment, the transition to a neighboring mode, quite probably the $\mathrm{TE}_{27,8}$ mode, around 137.5 GHz takes place in the beam voltage range $V_\mathrm{b} = 81.8$–$82.3\,\mathrm{kV}$. The simulations, however, continue in $\mathrm{TE}_{28,8}^{-}$ mode operation, and were extended up to $V_\mathrm{b} = 87\,\mathrm{kV}$, with $\alpha = 1.93$ with $\delta\alpha_\sigma = 13\%$.
The behavior of the codes for beam voltages exceeding the measured range was as follows:

- SELFT predicted a peak power of 1.23 MW at $V_\mathrm{b} = 84\,\mathrm{kV}$, and showed the rise of azimuthal neighbors and mode competition above this value, which reduced power to the range 650–750 kW.
- SELFT-BZ fitted better to the experiment, reaching a peak value exactly at the transition point $V_\mathrm{b} = 82.3\,\mathrm{kV}$. The diminished operation above, around 500 kW, was, however, due to strong pathological mode activity, and hence of limited significance.

[13] The absolute values of Δf discussed here are rather marginal, but since the deviations have a systematic appearance over the sweep, they might be relevant.

- EVRIDIKI, EVR-BZ and EVR-PIC performed very similarly, unfortunately by yielding powers up to the 1.4–1.5 MW range at the maximum simulated V_b, without severe mode competition or power reduction.

The very good match of frequency as well as power inside the measurement range does not help to explain the failure to reproduce the mode switch. A number of possible reasons for the different switching behavior are as follows:

- As previously mentioned, a weak parasite was present during the experiment. Although there is no basis for arguing that its existence significantly influences power or frequency below the beam voltage $V_b = 80\,\text{kV}$, it could of course influence the mode switch behavior in the upper parameter range.
- The external parameters, especially the beam voltage, are unrealistically steady in the simulation. V_b ripple in the experimental situation could in principle influence the stability in the high-power operation range.
- The pitch factor α above the transition range exceeds 1.5. Changes in the statistical distribution for high α (compare Section 2.1.1) could easily cause the treatment of the beam parameters to become inaccurate, overestimating α or underestimating $\delta\alpha_\sigma$.
- In the simulation scenario, the electron beam and the geometry are in perfect coaxial alignment. Eccentricity of the magnetic field axis or the cathode, even within the specified tolerances, could cause asymmetries which only start to affect the performance at the high-α operating points.
- The actual mode transition could be a multi-mode competition effect involving one ore several modes not in the list. Especially the many modes which are primarily relevant but had to be sorted out because of pathological effects are to be considered here.

- A simple possibility is also deviations of the gun simulation geometry from the technical implementation, regarding subtle geometry details and thermal deformations [Pag14].

Influence of the $\mathrm{TE}^{-}_{28,7}$ mode

Apart from the main sweep, data for the SELFT performance with the mode $\mathrm{TE}^{-}_{28,7}$ (SELFT^{+}), and of EVRIDIKI without it (EVR^{-}) are plotted in Figure 3.7.
Above $V_b = 75.3\,\mathrm{kV}$, the power in SELFT^{+} decreases rapidly, along with significant frequency deviations from the measured values. In these simulations, the $\mathrm{TE}^{-}_{28,7}$ mode appears to be frequency-locked to the nominal mode; this prevents direct experimental investigation of this phenomenon. The strong deviation from the measurement, however, indicates that this deterioration by the $\mathrm{TE}^{-}_{28,7}$ mode is not present in reality.
Interestingly, the EVR^{-}-simulation has different results than EVRIDIKI: Concerning both output power and frequency, it now follows the EVR-PIC curves almost exactly.
Except for SELFT^{+} above $76\,\mathrm{kV}$, all changes are within the confidence of the parameters, so with respect to agreement with the measurement both are equally true. If we, however, assume the EVR-PIC code to be the most trustworthy because it includes fewest physical simplifications, the following conclusion could be reached: The $\mathrm{TE}^{-}_{28,7}$ mode does not appear relevant in EVR-PIC; this simulation yields similar results as SELFT and SELFT-BZ, which had to be run without this mode because of numerical problems. As EVRIDIKI^{-} agrees very well with these results, the influence of the $\mathrm{TE}^{-}_{28,7}$ mode on EVRIDIKI and EVR-BZ might be as pathological as on SELFT^{+}, but only less pronounced.
Investigating such delicate differences between the different codes will certainly improve the understanding of the underlying differences in implementation; however, within this work, these paths are not further explored.

3.3.8 Benchmark summary

An overview short pulse (SP) sweep of a W7-X gyrotron over V_{cath} was presented and discussed, illustrating the general behavior of RF frequency and output power as a function of the applied cathode voltage. The experimental conditions for qualified frequency and power measurements were introduced, including a derivation of the corrections necessary when comparing SP calorimeter power measurements with simulation results. With the help of experimental long pulse power balance data and calorimeter specifications, a correction factor of $s_{\mathrm{corr}} \approx 1.080 \pm 0.02$ was determined.
The numerical simulation investigations enabled by this showed good agreement of the results calculated by several interaction simulation code variants with the experimental data, especially concerning the RF frequency with $|\Delta f| < 20\,\mathrm{MHz}$, i.e. $\pm 0.014\%$. It must be stressed that no analytical method was found which could reach a similar level of trueness[14]. In this context also the merit of swept investigations over single-operation-point studies becomes obvious, as the analytical methods' trueness was found to correlate roughly inversely with the RF power of the operating point. Comparison of the simulated RF powers also yielded satisfactory results over a large part of the sweep for all codes tested.
Unfortunately, major discrepancies between simulations and measurements were found concerning the mode transition in the cavity, which could not be reproduced properly. This leads to unrealistic power predictions in the high-V_{cath} section of the sweep, and also unfortunately drastically lowers the potential use of the mode transition position in the parameter space, e.g. for the investigation of parasitic oscillations and their impact on the operation.
Possible reasons for this deviation were listed; an elaborate investigation or even solution of this problem exceeds the scope of this work. The fact that all codes were affected in a similar way indicates that either a common modeling problem, for example interaction asymmetry

[14]The term is here used in the sense as it is defined by BIPM/ISO5725-1, as a measure for the systematic deviation of values from a reference [BIP08].

in the cavity, is the reason, or that the problem rises from the input data, most likely the beam parameters.
Rather technical problems in the self-consistent simulations were also addressed, concerning mainly the excitation of fast delocalized waves far from the electron cyclotron frequency, called "pathological modes" in this work. These underline the necessity for careful mode selection and attentive inspection of start-up processes to assure the physical consistence of a multi-mode simulation. The found varying susceptibility of the various codes to this phenomenon could be a first step to a better understanding of these effects.

4 Pulse spectrum analysis

In this Chapter, the developed pulse spectrum analysis (PSA) system and the underlying principles are presented, beginning with the hardware setup and the related consequences for dynamic range and bandwidth. Subsequently, the detailed operating principles of the measurement system are presented, and the focus is placed on the functionality and the robustness of the digital signal processing which carries out the RF reconstruction.[1] In addition, means to extract the information from the created spectrograms for post-processing and the possibility of extending the continuous acquisition length by undersampling are introduced.
The signal processing basics, especially A/D conversion, Fourier analysis and spectrograms, as well as the nomenclature for the related quantities, are compiled in Appendix A. To provide more space for detailed discussion, measurement examples are given in a separate, subsequent Chapter.

4.1 General hardware setup

A basic system setup with two hardware receiver channels is shown in Figure 4.1. A more comprehensive, but less illustrative full layout information on the system can be found in Appendix B, in combination with more technical information on the system.[2]

[1] The essential properties of the system can also be found in [SGJT13b]. A work-in-progress version was presented in [SGKT12]; however, it has to be noted that in this previous publication the RF reconstruction technique (see next Section) does not include the "cross sideband" R_x reconstruction and therefore postulates an inherent blind spot of the width Δf_d between the reconstructed HSI and LSI RF bands.

[2] The actual PSA system uses four physical channels with four mixers and four LOs, combined into two channel pairs. This allows the investigation of simultaneous events, even if they are separated by more than the span of one channel pair.

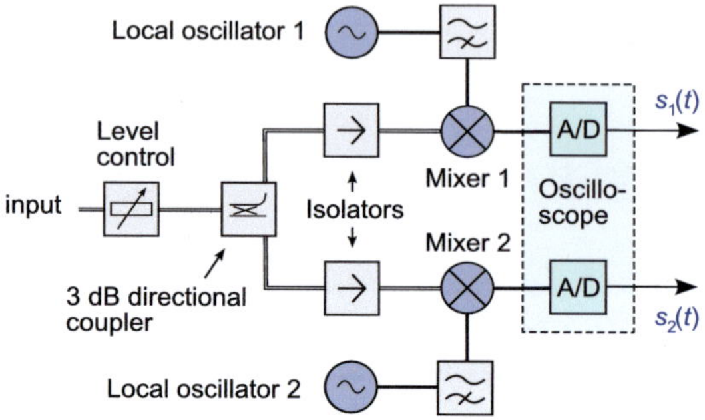

Figure 4.1: Hardware setup of the PSA system with one channel pair

The key components of the receiver front-end are two parallel harmonic mixers with conversion losses better than 40 dB for RF input signals in the range 110–170 GHz. Using the mixing harmonics of 10 or 12 requires local oscillator frequencies in the general range 9–17 GHz, allowing the use of standard synthesized sources operating at frequencies of 2–20 GHz. Since the mixers are driven by independent LO sources, their receiving range can be configured independently.[3]
For the A/D conversion process, a fast four-channel oscilloscope with a maximum sampling rate of $f_r = 10\,\mathrm{GS/s}$ is employed, yielding a Nyquist frequency of $f_N = 5\,\mathrm{GHz}$.[4] However, the specified bandwidth is 2 GHz; this is discussed in more depth in the following Section. The acquisition memory size of the device is 280 MS, which limits the gapless full-SR acquisition lengths t_a to 14 ms for two channels, and to 7 ms for simultaneous four-channel acquisition.
All data processing beyond the time domain A/D conversion is implemented in the PC-based measurement software, which coordinates the LOs and transfers the IF time domain signal data $s_1(t)$, $s_2(t)$ after acquisition to the measurement computer.[5]

[3]The setup in Figure 4.1 is *not* an inphase-quadrature receiver arrangement.
[4]In this work, we refer to amounts of samples in the dimensionless quantities GS (gigasamples), or MS (megasamples); a sampling rate (SR) of 10 GS/s hence means a sampling frequency of 10 GHz.
[5]Its structure and overall capabilities are summarized in Appendix B.

4.2 Considerations on dynamic range and bandwidth

Signals sampled with $f_N < f$ will appear aliased, with ambiguous frequency[6]. Since alias signal lines are a common problem in A/D converter based systems, the quantity *aliasing-free dynamic range* (AFDR) was defined, as the "safely" usable dynamic range without having additional knowledge about the incident signal.
In addition to this sampling-related problem, there are also impediments from other sources. Active microwave components or nonlinear elements inevitably create additional unwanted discrete frequencies, called spurious lines. In multi-channel systems, insufficient isolation between different signal branches can also cause this type of problem. In a similar fashion as the AFDR, a *spurious-free dynamic range* (SFDR) is defined. Increasing the acquisition length t_a reduces the continuous background noise level, but does not affect the spurious lines and alias product magnitudes. In a broadband measurement system, the safely usable dynamic range will therefore be obtained when the SNR is roughly equal to, but not exceed the maximum SFDR and AFDR level: This provides a continuous noise floor without undesired discrete sigals prone to misinterpretation.
In the following, first the capabilities of the receiver in the PSA system are discussed, then the already described RF front end is also taken into account. After the description of the complete SFDR and AFDR, the sensible parameter values for the FFT length are discussed.

4.2.1 IF receiver

Several properties of the oscilloscope acting as an IF receiver have a direct impact on the achievable total dynamic range:

- channel isolation
- input low-pass edge frequency and selectivity
- jitter based degradations (expressed as effective number of bits, ENOB, compare Appendix A.2)

[6]Compare derivations and discussions in Appendix A.1.

If several channels are used simultaneously, the channel isolation must be taken into account, which is specified to exceed 60 dB for $f < 2\,\mathrm{GHz}$ and 50 dB for $f > 2\,\mathrm{GHz}$. The front-end low pass filter curve with its edge frequency f_{LP} determines the IF flatness, and also the robustness against aliasing. The combination of the two properties defines the AFDR. Signals with $f > f_{\mathrm{LP}}$ will still appear in the IF range, but will be attenuated by the low pass.
A measured spectrum for a sweep of the input signal in 0–10 GHz is given in Figure 4.2, which shows directly the shape of the input low pass with its 3 dB point at 2 GHz and a pole at f_{N}. In the figure also the mixer diplexer was included, which separates LO and IF. Since its IF branch is a low pass, aliasing are further reduced by it. The combined signal of the oscilloscope low pass and the diplexer is sampled with 10 GS/s, which means the range $5 < f/\mathrm{GHz} < 10$ appears mirrored in the first Nyquist zone. Hence, the difference between the true first-zone signal and the alias directly shows the AFDR.
The specified minimum isolation as noted above was also inserted in zone 1, as a down-shifted copy of the "Combination" curve. Together with the mirrored aliasing-curve, it indicates directly which part of the IF range is primarily alias- and which is isolation-spurious limited.[7] Since the main internal spurious source is channel isolation, the SFDR can be read directly from the figure.
Since the timing jitter of an A/D-converter influences the usable combinations of dynamic range and bandwidth (compare Appendix A.2), the oscilloscope employed as the receiver was chosen for its single-unit A/D converter with optimized jitter-limited dynamic range [Roh14].[8] The measured ENOB of the device is 7.2 for $f < 2\,\mathrm{GHz}$, and decreases because of the input low pass slope for higher frequencies.

[7] This is a conservative estimate, as only the alias curve is measured, while the isolation curve respresents a specified minimum value. Test measurements with the device, which, however, cannot cover all the possible frequency and settings combinations, have yielded isolations better than 70 dB in the whole frequency range.

[8] In typical fast oscilloscopes, n A/D units with identical sampling rate f_{r} are implemented in parallel, detecting the incoming signal in an interleaved fashion. This yields a combined SR of nf_{r}, but increases jitter problems because of synchronization limitations between the units.

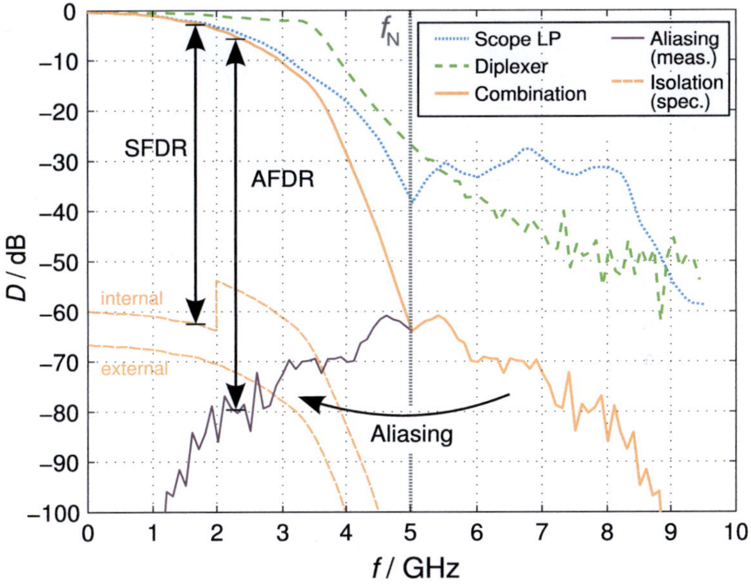

Figure 4.2: Spurious and alias signal measurement at the oscilloscope

4.2.2 Mixers and local oscillators

From the mixer front-end, two further possible impediments for an otherwise well-configured receiver can rise. The first is the generation of spurious lines, which can be misinterpreted as valid down-converted RF signals if visible in the IF range. The sources of these spurious lines can be separated as follows:

- RF mixer crosstalk can occur over the junction which separates the RF input path into the mixer channels (compare Fig. 4.1). Here, the up-converted f_1^{LO} can be radiated through mixer 1 to be received in channel 2 like a regular RF signal. This effect can be mitigated through hardware-implemented channel isolation, by employing a directional coupler in combination with isolators at each mixer. In the current setup, this is achieved through a

3 dB coupler with an isolation of better than 27 dB in combination with full-band isolators, providing additional isolation of −20 dB at the RF port of each mixer. Assuming that in a worst case the LO is up-converted and radiated by mixer 1 with the same conversion loss as typically experienced in normal operation, i.e. with > 30 dB, the maximum LO-caused IF spurious levels over the RF path can be estimated as:

+10 dBm	LO drive power
−30 dB	up-conversion in mixer 1
−20 dB	isolator
−27 dB	coupler
−30 dB	down-conversion in mixer 2
−97 dBm	IF power at receiver

- Another source for IF spurious lines is direct LO-IF crosstalk. Many synthesized sources create significant spurious signals, which for example can appear at fractional frequencies of the current f^{LO}, and thus within the receiver frequency range. If these spurious lines are able to penetrate with levels above the IF noise floor, they cannot easily be distinguished from valid signals.
  Network analyzer measurements found that the mixer diplexer's LO/IF isolation varies in the range 10–50 dB for $0 < f/\mathrm{GHz} \leq 3$, and even falls below 10 dB between 3 and 4 GHz. With the specified LO fractional spurious levels up to −60 dBc, i.e. −50 dBm, this not sufficient to protect the system. Additional high-pass filters with a minimum LO-IF rejection of 67 dB in $f < 6.5\,\mathrm{GHz}$ were added to the LO path, ensuring direct LO spurious levels in the IF range to be below −117 dBm.

The second possible problem is the breakthrough of LO noise into the IF range. This can either be strong noise in the IF range, which directly propagates through the mixer into the receiver IF path, or noise around f^{LO} which is then converted by the mixer and appears the in the vicinity of the RF position of the LO signal $f^{\mathrm{RF}} = h \cdot f^{\mathrm{LO}}$. While the former problem is alleviated by the LO high pass filter mentioned above, the latter is a fixed property of the signal source.

4.2.3 Total dynamic range

The previous considerations yield an alias- and spurious-free dynamic range $D = 60\,\mathrm{dB}$ for $f < 2\,\mathrm{GHz}$ and $D = 50\,\mathrm{dB}$ for $2 < f/\mathrm{GHz} < 3.7$, entirely dominated by the maximum isolation specification. For higher frequencies the AFDR is smaller than the specified isolation SFDR, and naturally reaches zero for $f \to f_{\mathrm{N}}$.
If we adjust the internal isolation limitation to the minimum measured value of $72\,\mathrm{dB}$, the next limitation encountered is the external isolation of at least $67\,\mathrm{dB}$. With this criterion the attainable aliasing- and spurious-free dynamic range is at least $D = 60\,\mathrm{dB}$ for $f^{\mathrm{IF}} \leq 3\,\mathrm{GHz}$, corresponding to an identical IF bandwidth $\Delta f_{\mathrm{IF}} = 3\,\mathrm{GHz}$.
To establish the total frequency domain dynamic range as derived in Appendix A.2, the load factor in (A.15) must be defined. This can be done in a straightforward way by estimating the maximum power at the oscilloscope input over the mixer properties. The typical wideband conversion loss C is in the range of $30\,\mathrm{dB}$, and with the $1\,\mathrm{dB}$ compression point at $P_{\mathrm{RF,in}} = -10\,\mathrm{dBm}$ this yields a maximum IF power of $-40\,\mathrm{dBm}$, or $V_{\mathrm{s}} = 3.16\,\mathrm{mV}$. With the most sensitive vertical detection range of $\pm 5\,\mathrm{mV}$, this leads to a linear load factor of $L = 0.63$.[9]
The time domain discretization SNR can be hence calculated to be equal to $SNR_{\mathcal{T}} = 24.1\,\mathrm{dB}$. As stated before, a consequent way to define a maximum default STFT resolution $\delta f = 1/t_{\mathrm{seg}}$ is to find the threshold of possible aliasing/spurious emergence from the continuous background noise, using (A.16) as

$$SNR_{\mathcal{F}} = SNR_{\mathcal{T}} + 10\log_{10}\frac{N}{2} \leq D \quad . \tag{4.1}$$

[9] The front-end includes a wide-band automatic level control loop which was already mentioned in Section 3.1.2. This allows the use of optimal load factors within a power range of ~30 dB, further enhancing the overall dynamic range of the system.

With $f_\mathrm{r} = 10\,\mathrm{GHz} = 1/t_\mathrm{r}$ and $t_\mathrm{a} = N \cdot t_\mathrm{r}$, this equation can be solved for the maximum DFT time segment length which is guaranteed not to expose undesired spurious or aliasing signals:

$$t_\mathrm{a} \geq \frac{2}{f_\mathrm{r}} \cdot 10^{0.1(D-SNR_\mathcal{T})} = 0.770\,\mathrm{ns} \approx 1\,\mathrm{\mu s} \quad . \tag{4.2}$$

Of course, it is justified to use higher frequency resolutions, e.g. for line width estimations or modulation analysis. It should only be kept in mind that additional weak lines might not be valid signals.

4.3 IF data processing

The detected IF data consists of the two independent time signals $s_1(t)$ and $s_2(t)$. By application of the STFT to these individual signals $s_i(t)$, the IF spectrograms $S_i(f,t)$, with frequency axis f and time axis t, are generated. This process is derived and discussed in Appendix A.3. Both signals are processed with the same STFT settings, resulting in identical numerical resolutions δf and δt for both of the S_i.
The spectrogram sizes in terms of time bins n_t and frequency bins n_f are given by

$$n_\mathrm{f} = \frac{\Delta f_\mathrm{IF}}{\delta f} \tag{4.3}$$

$$n_\mathrm{t} = \frac{t_\mathrm{a}}{(1 - x_\mathrm{OV})\, t_\mathrm{seg}} \quad , \tag{4.4}$$

with $t_\mathrm{seg} = 1/\delta f$. For the typical settings $\Delta f_\mathrm{IF} = 3\,\mathrm{GHz}$, $t_\mathrm{a} = 4\,\mathrm{ms}$, $t_\mathrm{seg} = 1\,\mathrm{\mu s}$ and $x_\mathrm{OV} = 0.2$, the number of values along the two axes are $n_\mathrm{f} = 3000$ and $n_\mathrm{t} = 5000$. The size of the resulting discrete spectrograms is therefore $n_\mathrm{f} \cdot n_\mathrm{t} = 15$ Megapixels.[10]

[10] Increased x_OV are indicated when a demand for high frequency resolution drives the order of magnitude of t_seg close to the one of t_a, as then "trivial" spectrograms with small n_t values occur. This means that if in the above example $\delta f = 20\,\mathrm{kHz}$ was required, the result dimensions would be $n_\mathrm{t} = 100$ and $n_\mathrm{t} = 150 \cdot 10^3$; apart from the large number of frequency points, this spectrogram has an impractical aspect ratio. With $x_\mathrm{OV} = 0.95$, this is alleviated to $n_\mathrm{t} = 1600$, which still gives an impression of time dependence; the cost is increased processing and memory load.

These IF spectrograms are still subject to the mixing ambiguities discussed in Section 1.3.1, and therefore of only limited use for gyrotron spectrum analysis. However, they already display the measured time-variant spectrum.
The frequency range of these spectrograms is defined by the local oscillator frequencies f_i^{LO} at acquisition time. Since the individual channel signals s_i were received simultaneously, the spectra $S_1(f, t_1)$ and $S_2(f, t_2)$ describe simultaneous properties for $t_1 = t_2$. All following operations are only applied to such simultaneous spectra.

4.4 RF spectrum reconstruction

As discussed in Section 1.3.1, all signals received through the available mixing harmonics and injection sides in a broad-band superheterodyne will be situated in the IF range. When attempting to determine the original RF position of frequency components found in the IF spectrograms under discussion, one faces substantial ambiguity since neither the mixer injection side nor the acting harmonic is known. Also, the individual traces found in the IF range are generally received through different mixing harmonics and injection sides.
To avoid the necessity of a priori knowledge about the RF spectrum or subsequent validation measurements, the redundancy of the twin receiver channels is exploited. In this process, the signals of the two channels are merged, because of which we refer to these as a receiver *channel pair*.

4.4.1 Principle

In a first step, a *desired acting harmonic* $h_{n,\mathrm{d}}$ for each channel n is defined. The goal of the processing is then to eliminate all signals from both channels which have not been received through $h_{n,\mathrm{d}}$, and to separate the remaining signals by the injection side. We refer to this process as *RF reconstruction* in this work, as the final result is a representation of the RF spectrum.

The key to the reconstruction is the introduction of a specific *desired LO frequency difference*

$$\Delta f_{\mathrm{d}} = h_{1,\mathrm{d}} \cdot f_1^{\mathrm{LO}} - h_{2,\mathrm{d}} \cdot f_2^{\mathrm{LO}} \tag{4.5}$$

on reception. For nontrivial choices $f_1^{\mathrm{LO}} \neq f_2^{\mathrm{LO}}$ the RF signals received in the IF spectrograms appear effectively shifted with respect to each other in the IF range.[11]
The amount of relative frequency shift depends on the acting mixer harmonic h_n, while the direction of the shift can separate low and high side injection. This is illustrated in Figure 4.3a, where three RF signals marked by "□", "Δ" and "∇" are received simultaneously in the two channels. The incident RF signals (on the right) are downconverted with the effective LO frequencies $h_{\mathrm{d},1} \cdot f_1^{\mathrm{LO}}$ and $h_{\mathrm{d},2} \cdot f_2^{\mathrm{LO}}$. Note how the frequency difference between the LO frequencies causes the received IF signals (left side) to be located in different frequency positions.
To the measurement system and the user only these IF signals are known, and are used to reconstruct the original RF spectrum. For a numerical example, consider a monofrequent signal in IF channel 1 at the frequency $f_1^{\mathrm{IF}} = 2\,\mathrm{GHz}$, with $f_1^{\mathrm{LO}} = 15.0\,\mathrm{GHz}$. The ambiguity in this case was already described in Section 1.3.1, what follows is the technique for its removal. For a configuration $h_{1,\mathrm{d}} = 10$, $h_{2,\mathrm{d}} = 12$, $\Delta f_{\mathrm{d}} = 200\,\mathrm{MHz}$, with the help of (4.5), the channel 2 LO has been accordingly set to be equal to

$$f_2^{\mathrm{LO}} = \left(h_{1,\mathrm{d}} \cdot f_1^{\mathrm{LO}} - \Delta f_{\mathrm{d}}\right) \cdot 1/h_{2,\mathrm{d}} = 12.48\overline{3}\,\mathrm{GHz} \tag{4.6}$$

on reception. The received signals can then be related as follows: If the signal at f_1^{IF} was received through $h_{1,\mathrm{d}} = 10$ LSI, this indicates that an original RF of

$$f^{\mathrm{RF}} = h_{1,\mathrm{d}} \cdot f_1^{\mathrm{LO}} + f_1^{\mathrm{IF}} = 152\,\mathrm{GHz}$$

is represented by f_1^{IF}. Then, in channel 2 a corresponding signal at

$$f_2^{\mathrm{IF}} = f^{\mathrm{RF}} - h_{2,\mathrm{d}} \cdot f_2^{\mathrm{LO}} = 2.2\,\mathrm{GHz}$$

[11] In the following we refer only to this definition, which is arbitrary with respect to the relative RF position of the $h \cdot f^{\mathrm{LO}}$; it could alternatively be made inverse as $\Delta f_{\mathrm{d}}' = h_{2,\mathrm{d}} \cdot f_2^{\mathrm{LO}} - h_{1,\mathrm{d}} \cdot f_1^{\mathrm{LO}}$, without changing the workings of the technique.

must be present. A non-existence of this validating channel 2 IF signal excludes this reception possibility, and indicates a different reception harmonic-injection side combination.
Next, the HSI reception case with the same harmonic $h_{1,\mathrm{d}} = 10$ could be checked, yielding

$$f^{\mathrm{RF}} = h_{1,\mathrm{d}} \cdot f_1^{\mathrm{LO}} + f_1^{\mathrm{IF}} = 148\,\mathrm{GHz} \quad .$$

Here, a validation signal at $f_2^{\mathrm{IF}} = 1.8\,\mathrm{GHz}$ is postulated. If both described validation signals cannot be found, the signal was not received through the desired harmonics and can be discarded; of course further harmonic combinations also can be checked in the same way.

4.4.2 Implementation

The example above was formulated for signals of defined frequency; however, the available data consists of spectrograms with fixed frequency bins. Considering the typical size of an IF spectrogram with a total number of frequency bins in the order of millions, the validation operation must be applicable on a bin-by-bin basis in an efficiently automatized way.
A simple and straightforward implementation is the application of the *min()*-operation on numerically shifted and mirrored instances of the spectra $S_1(f, t_i)$ and $S_2(f \pm \Delta f_\mathrm{d}, t_i)$, instead of looking for signals and calculating their counterpart in the other channel.
Performing this comparison through shifting and merging with *min()* yields with

$$R_{\mathrm{LSI}} = \mathrm{Min}\left(S_1(f, t_i), S_2(f - \Delta f_\mathrm{d}, t_i)\right) \quad \forall\, i \tag{4.7}$$

$$R_{\mathrm{HSI}} = \mathrm{Min}\left(S_1^{\leftrightarrow}(f - \Delta f_\mathrm{d}, t_i), S_2^{\leftrightarrow}(f, t_i)\right) \forall\, i \tag{4.8}$$

the reconstructed RF spectrograms received in both mixer sidebands for the selected mixer harmonic in a primitive, yet effective way. It must be noted that this operation only relies on the independent comparison of f-t bin pairs, and thus can be performed parallelized.

The described process is illustrated in Figure 4.3b for the three signals introduced in Figure 4.3a. Note how for the LSI reconstruction the two IF spectra are shifted by Δf_{d} against each other; this causes the signals labeled "Δ" to have the same IF frequency; the following *min()* on the two spectra thus removes all signals except this one.[12]
The HSI reconstruction works analog, except for the additional mirroring of the IF spectra by the frequency $0\,\mathrm{Hz}$. In this case, the signal labeled "∇" is reconstructed on the left side of Figure 4.3b in agreement with the fact of its reception in HSI in both channels, as visible in the RF part of Figure 4.3a.
As demonstrated above, the explained technique as formulated in (4.7) and (4.8) allows the reconstruction the RF range for the overlapping range of the two channel's coverage, separately in LSI and HSI. Of course, the total achievable bandwidth is reduced by the fringe part of the width Δf_{d} at both shifted IF band edges, where no validation data from the other channel are available.
There is another, also yet unused part of the IF spectrum, which can still be evaluated: $0 \ldots \Delta f_{\mathrm{d}}$, the part of the spectrum which was received with HSI in channel 1 and LSI in channel 2. It can be recovered by

$$R_{\mathrm{x}} = \mathrm{Min}\left(S_1^{\leftrightarrow}(f + \Delta f_{\mathrm{d}}, t_i), S_2(f, t_i)\right) \quad \forall\, i \tag{4.9}$$

in a similar way. This removes the blind spot which was postulated for this reconstruction method in early publications [SGKT12]. In Figure 4.3, this type of signal is represented by the component labeled with "$\square$".
As can be seen in Figure 4.3, the reconstructed spectra R_{HSI}, R_{X} and R_{LSI} can be recombined to form a fully reconstructed spectrum.[13] The effective bandwidth of the total reconstructed RF spectrum is equal to $2\Delta f_{\mathrm{IF}} - \Delta f_{\mathrm{d}}$ if all usable IF range is exploited.

[12] As in the Figure the relation between Δf_{d} and Δf^{IF} is exaggerated, it is also visible how the bandwidth of this spectrum is necessarily reduced from Δf^{IF} to $\Delta f^{\mathrm{IF}} - \Delta f_{\mathrm{d}}$ in this operation.

[13] Compare the final result of the reconstruction process as shown in the bottom part of Figure 4.3b with the incident RF spectra in Figure 4.3a.

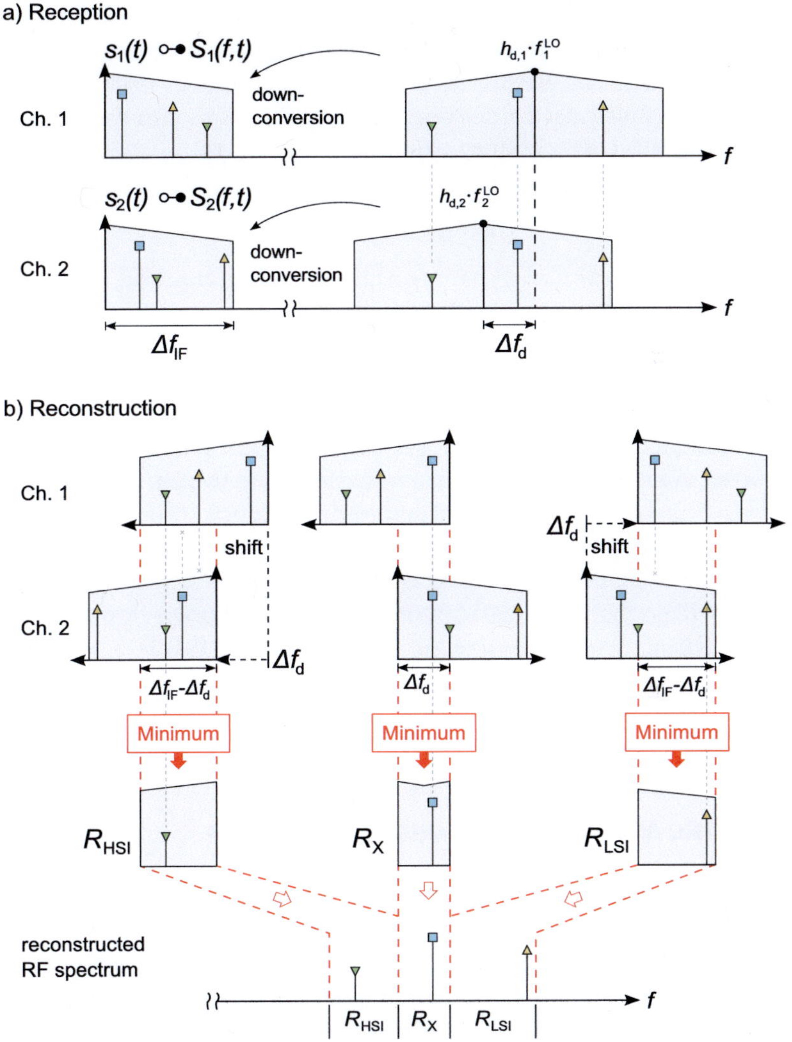

Figure 4.3: Reception of multiple signals through channel pair (a) and reconstruction of RF spectrum (b) using mirrored and frequency-shifted IF spectra

A basic advantage of this reconstruction technique is the combination of a highly nonlinear decision function without an increase of result complexity. The result still has the convenient power unit dBm, and no detection threshold values or weighting functions and the like have to be considered when interpreting it.

4.4.3 Effective dynamic range

If there is level imbalance between the signals compared, the highly nonlinear operation *min()* will cause the result spectrogram R to have a different dynamic range than the input data S_i.
In the case of flat imbalance, i.e. one receiver channel generally has lower signal levels possibly due to higher mixer conversion loss, the reconstruction will simply adopt the dynamic range of the weaker channel. Weak signals which are only visible in the stronger channel will be removed by the reconstruction process.
Frequency-dependent IF imbalance however can have a more visible effect on the result, depending on Δf_d: Uneven conversion performance within the IF bandwidth or a pronounced noise floor from the LO can form a frequency-variant signal background. If Δf_d is of the same order or larger than the frequency distance between differently shaped areas of the IF noise background, valid weak signals could be masked, and invalid weak signals could persist.

4.5 Reconstruction robustness criteria

The described reconstruction aims at the removal of frequency ambiguities; robustness here means the reliability of the result.
RF ambiguity generally arises when a signal appears in the reconstructed spectrogram at a frequency where it was not actually present on reception. This happens when the relative frequency difference Δf between two signals in S_1 and S_2 satisfies $|\Delta f_\mathrm{d} - \Delta f| < \delta f$, while the two signals have not been received through the designated harmonics. We refer to such signals as false-positive validations for which the two principal sources are separated and discussed in the following.

4.5.1 Coincidental false-positive validations

This case describes the reception of two arbitrary and unrelated spectral components in a way that they satisfy the reconstruction condition. If the components of these independent signals coincide in IF with Δf_{d}, they mingle indistinguishably and lead to a falsely reconstructed frequency component in R. This could, for example, happen for an unfortunate combination of different oscillations in the gyrotron (for example, a parasitic oscillation and a cavity oscillation) or the combination of a IF static spurious line with the mixing product of a genuine RF signal.
It must be noted that static spurious signals have no possibility to appear with the expected frequency shift in the different IF channels and are removed in the reconstruction process. Of course, they still should be minimized because they increase the probability of the aforementioned frequency coincidence leading to false-positive validations. Since the technique has no inherent safeguards against coincidental false-positive reconstruction, it is obvious that only sparse (i.e. not broadband / continuous) spectra can be treated with a low risk of this error.

4.5.2 Deterministic false-positive validations

There is a number of sources for unwanted frequency components in the IF channels, which can exhibit a relation of Δf_{d} per definition. This would e.g. be the case for spurious lines from the LOs, if they had a fixed frequency offset relationship to the set LO frequency.
Alias signals are a special case; alias reception in parallel channels with slightly shifted RF LO positions can be separated in three different basic ways, depending on whether the compared signals reside in the same Nyquist zone.

1. $2f_{\mathrm{N}} > f_1^{\mathrm{IF}} > f_{\mathrm{N}} > f_2^{\mathrm{IF}}$ describes the "mixed" reception case, where the channel 1 signal is aliased while the channel 2 signal is sampled correctly. The frequency difference between the two signals is not Δf_{d} because f_1 was mirrored at f_N, and is received

as $f_1' = 2f_\mathrm{N} - f_1$ (compare eq. A.8). This leads to a removal of both signals during the reconstruction process, as in all regions where the two signals do not reside in the same Nyquist zone.

2. With $2f_\mathrm{N} > f_1^\mathrm{IF} > f_2^\mathrm{IF} > f_\mathrm{N}$, both signals are in the second Nyquist zone. As both signals are mirrored at f_N, the frequency difference $\Delta f = -\Delta f_\mathrm{d}$ satisfies the reconstruction condition. However, because of the sign change, the signals will be reconstructed in the wrong injection side and thus at the wrong RF.

3. If with $3f_\mathrm{N} > f_1^\mathrm{IF} > f_2^\mathrm{IF} > 2f_\mathrm{N}$ both signals are in the 3rd Nyquist zone, a similar effect occurs. With $\Delta f = \Delta f_\mathrm{d}$, the correct injection side will be found, but the resulting RF will remain wrong.

These effects are the reason for the relevance of the AFDR for the usable dynamic range in Section 4.1. The occurrence of aliasing cannot be detected from the measurement data without additional knowledge.

4.5.3 Estimation of reception possibilities

Apart from the challenges discussed before, there remains the investigation of the possibility of receiving a valid signal through undesired harmonics, leading to a false-positive detection. Unlike the sources of false validations discussed before, this risk is heavily dependent on the choice of $h_{n,\mathrm{d}}$. While the STFT parameters can be chosen after signal detection, the LO frequencies and in the same decision the mixing harmonics and Δf_d must be chosen before. Thus, a minimization of the deterministic false-positive validations, which cannot be ruled out by design, is investigated here.
As stated before, a false-positive validation occurs when in both channels the down-converted signal is located inside the IF bandwidth, with a frequency difference $\Delta f \approx \Delta f_\mathrm{d}$. For a given RF range of a possible incident signal and a set of chosen receiving harmonics $h_{n,\mathrm{d}}$, all possible combinations of probable other acting harmonics h_n and the resulting IF position of the received frequency can be calculated. In order to assess different configurations $h_{n,\mathrm{d}}$, it is sufficient to evaluate

how close a reception through unwanted harmonics can get to the reconstruction criterion in a specified RF range of possible signals. This can be expressed as a frequency difference ratio

$$F = |\frac{\Delta f}{\Delta f_{\mathrm{d}}} - 1| \tag{4.10}$$

where Δf is the occurring frequency difference and Δf_{d} is the value expected for validation.
Relevant technical F ranges can be separated as follows:

1. $F = 0$ means mathematical identity of the observed frequency difference with the expected value. If this occurs through an unfortunate combination of harmonics and signals for $h_n \neq h_{n,\mathrm{d}}$, a false reconstruction is inevitable.

2. $0 < F < 2\delta f/\Delta f_{\mathrm{d}}$ indicates a frequency shift deviating from Δf_{d} which is too small to be resolved by the system. This could be countered by choosing a higher Δf_{d} value (only a-priori) or by increasing δf (also a-posteriori).

3. $2\delta f/\Delta f_{\mathrm{d}} \leq F < \Delta f_{\mathrm{IF}}/\Delta f_{\mathrm{d}}$ yields lines which are filtered by the reconstruction technique as intended and therefore basically do not generate false-positives. It should however be kept in mind that the linewidth of the signal itself should also be taken into account, so an F value further away from the lower border will provide a safety buffer.

4. $F \gtrsim \Delta f_{\mathrm{IF}}/\Delta f_{\mathrm{d}}$ denotes the case where one of the two received signals per definition does not lie in the receiver IF range, and therefore is discarded by the reconstruction. This is the case of maximum robustness.

A given choice of $h_{n,\mathrm{d}}$ can be characterized by choosing an RF input range and then finding the minimum F which can be caused by an arbitrary frequency in this RF range, letting F convey the worst-case reception through undesired harmonics, with respect to robustness.

With the WR6 waveguide lower cut-off frequency and twice the intended maximum RF, the possible RF range 90–340 GHz is suitable for this investigation. A graphical representation of the resulting minimum F values can be found in Figure 4.4; because of the wide range of possible F values, here first all results with $F > 1$ were set $F := 1$ to give a good indication of the smallest F values.

A first positive result is that no combinations are found which yield $F = 0$ (case 1 in the list), so the basic reconstruction is never violated, and robustness for a given set of $h_{n,\mathrm{d}}$ is a matter of proper δf and Δf_d choice. For a typical configuration with $\delta f = 1\,\mathrm{MHz}$, $\Delta f_\mathrm{d} = 100\,\mathrm{MHz}$ and $\Delta f_\mathrm{IF} = 3\,\mathrm{GHz}$, the upper border of the case 2 listed above is $2\delta f / \Delta f_\mathrm{d} = 0.02$, which also does not occur within the range considered. Case 3 extends up to $\Delta f_\mathrm{IF} / \Delta f_\mathrm{d} = 30$, which applies for practically all values in this calculation. This case hence is the one most relevant

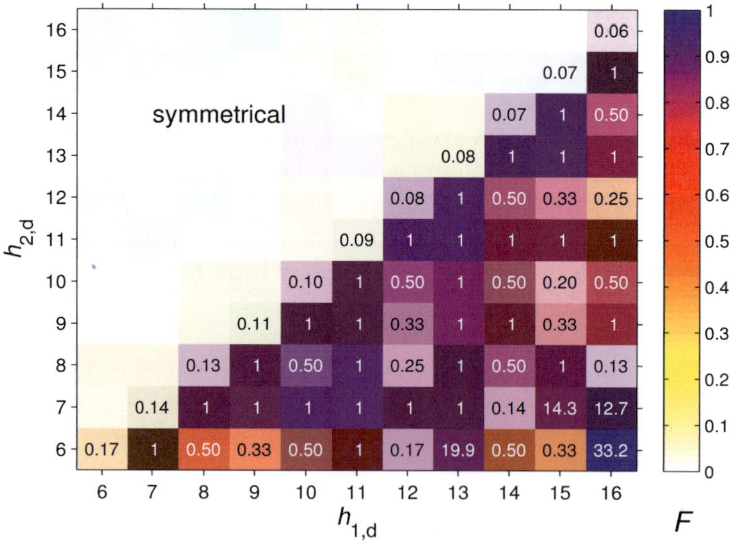

Figure 4.4: F for worst-case Δf results using fixed sets $h_{n,\mathrm{d}}$ (axes) with $f^\mathrm{RF} = 90$–$340\,\mathrm{GHz}$; $F >$ results are displayed as $F = 1$

for the assessment. The maximum robustness case is only found for $h_{1,\mathrm{d}} = 16$, $h_{2,\mathrm{d}} = 6$ in the lower right corner.
The small robustness of $h_{1,\mathrm{d}} = h_{2,\mathrm{d}}$ is obvious, which stems from the reception through neighboring harmonics, i.e. $h_n = h_{n,\mathrm{d}} \pm 1$. Interestingly, for most harmonics the value $F \geq 1$ is obtained. The harmonics which cause this are multiples and fractions of the h_d, mostly $h_n = 2h_{n,\mathrm{d}}$ or $h_n = h_{n,\mathrm{d}}/2$, so the hypothetical signal necessary for this case to occur is far away from the signal of interest.
It is the choice of 90–340 GHz for the potential RF range which causes these cases to occur, often on basis of signals far away from the actually used bandwidth. If in a more lax estimation only the technical range of 110–170 GHz is chosen, considerably more optimistic robustness values are estimated as can be seen from Figure 4.5; here the F range was not limited.

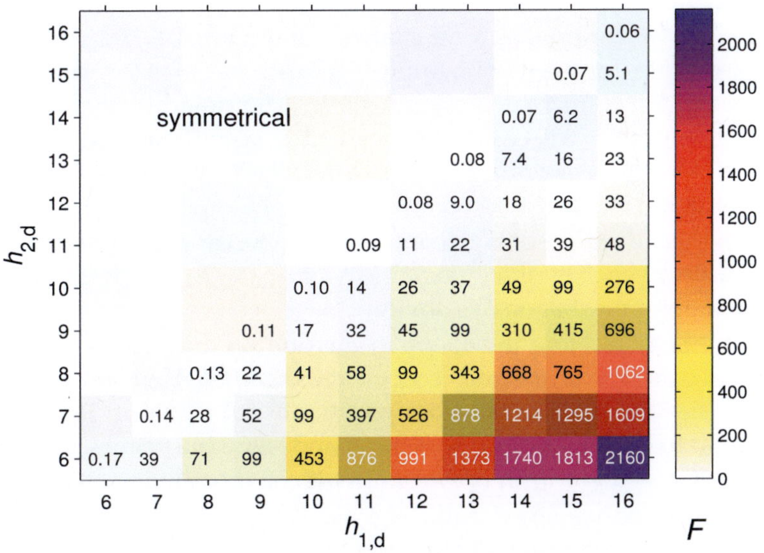

Figure 4.5: F for worst-case Δf results using fixed sets $h_{n,\mathrm{d}}$ (axes) with $f^{\mathrm{RF}} = 110$–$170\,\mathrm{GHz}$; full F range is plotted

Clearly for many harmonics combinations now F values of the maximum robustness category $F \gtrsim 30$ are reached.
However, in both estimations it is obvious that the choice of $h_{n,\mathrm{d}}$ has a noticeable impact on the constraints for the chosen configuration; the higher the F value found for a given harmonic pair, the less critical are the other parameters. For the calculations shown in Figure 4.4 it can be stated that the minimum F is dependent on the lowest common denominator of the chosen harmonics, and decreases weakly with rising harmonic value. It is noticeable that consequently, peak performance is delivered by all nonidentical combinations of prime mixing harmonics.
As $F \geq 1$ means that a signal of the bandwidth Δf_d can be reconstructed properly, it is worth investigating how many problematic harmonic combinations which cause $F < 1$ exist for each set $h_{n,\mathrm{d}}$. Minimizing this total number N of possible unwanted reception cases contributes to robustness in an additional way.
In Figure 4.6, for each $h_{n,\mathrm{d}}$ set analyzed in Figure 4.4, N is displayed. For example, for the combination $h_{1,\mathrm{d}} = 12$, $h_{2,\mathrm{d}} = 9$ (or vice versa), there are $N = 3$ reception cases $h_i \neq h_{n,\mathrm{d}}$ for an arbitrary RF signal in the range $90 - 340\,\mathrm{GHz}$ to be received in both IF ranges with an F value below 1. According to Figure 4.4, the most critical of these 3 combination has $\Delta f = 0.77\Delta f_\mathrm{d}$ or $\Delta f = 1.33\Delta f_\mathrm{d}$.
Again the advantage of high common denominators between the $h_{n,\mathrm{d}}$ is visible. Especially the vulnerability of $h_{1,\mathrm{d}} = h_{2,\mathrm{d}}$ with the high N values along the diagonal is obvious.
It must be noted that the above discussion on the basis of the three figures concerns undesired reception *possibilities*. They are of course only relevant if there exists an additional RF signal in the proper position to be received through these undesired harmonics.
Since the even gyrotron operation with considerable parasitic activity features only a few parallel signals, it can be assumed that these investigations are rather conservative.[14]

[14]Note, however, that modulation can impose problems if Δf is chosen smaller than the modulation bandwidth.

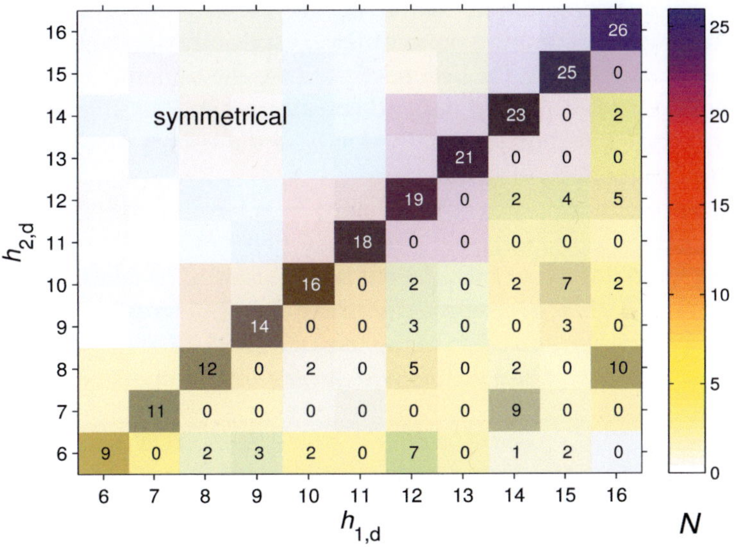

Figure 4.6: Number N of undesired reception possibilities in the frequency range $f^{RF} = 90\text{–}340\,\text{GHz}$ (compare also Fig. 4.4), i.e. sets of $h_n \neq h_{n,d}$, with $F < 1$

4.6 Summary on parameter choices

4.6.1 Desired mixing harmonics

While the choice $h_{1,d} = h_{2,d}$ has an intuitive appeal because it allows the use of identical mixers and promises similar frequency response in both channels, it provides the smallest robustness against RF ambiguity. This is obvious from inspecting Figures 4.4 and 4.6; e.g. a choice of $h_{n,d} = 12$ leads to 19 possible harmonic reception cases with $F < 1$.[15] An obvious better choice for robustness are sets with $h_{1,d} \neq h_{2,d}$. For most of these configurations, smaller values of $\Delta f_d < 100\,\text{MHz}$ are

[15]Since $F = 0.08$, for $\Delta f = 100\,\text{MHz}$, a frequency resolution better than 8 MHz would in principle still be able to remove these signals in the reconstruction process.

not problematic, as can be seen in Figure 4.4. Generally, the choice of sets with a large common denominator, especially $h_{n,\mathrm{d}}$ sets involving prime numbers, yield the best results. Since the dynamic range of the system on the other hand depends on similar and flat IF response of the mixers, such choices usually decrease the usable dynamic range through conversion loss imbalance.
The choice of harmonics is also determined by the mixers employed. In the current setup, wide-band even-harmonic mixers are used, which of course do not allow the efficient use of uneven or relatively prime harmonics. A good compromise between the needs for dynamic range, bandwidth and robustness is the use of $h_{1,\mathrm{d}} = 12$, $h_{2,\mathrm{d}} = 10$, which was found to yield good results with a very small impact on the usable dynamic range.

4.6.2 LO frequency shift Δf_d

Apart from the importance for reconstruction robustness explained above, the value of Δf_d has two important impacts:

1. With a large shift, the frequency responses of the mixers will differ. Due to the *min()*-operation, conversion loss imbalance will reduce the dynamic range of the system, and lead to reconstruction artifacts caused by the uneven noise floor, as discussed before.

2. Furthermore, the usable IF bandwidth is reduced by Δf_d, as the total frequency overlap between the IF spectrograms is decreases with increasing Δf_d.

While these two criteria require a minimization of Δf_d, the reconstruction robustness investigation results call for large values. Also, the maximum allowable continuous bandwidth of a measured signal for the process to retain the described unambiguity without an additional manual check on the IF spectrograms also directly relates to the LO frequency difference. For typical operation of the system, values of $\Delta f_\mathrm{d} = 50\text{–}300\,\mathrm{MHz}$ are reasonable.

4.6.3 FFT / STFT configuration

For this RF reconstruction technique to work properly, low window side-lobes and good side-lobe roll-off are desirable, while the main lobe width is of lesser importance. For this reason, the Blackman-Harris window was chosen as an appropriate window function. It yields $67\,\mathrm{dB}$ side-lobe suppression and a $3\,\mathrm{dB}$ width of 1.66 frequency bins [Har78].
A practical problem already mentioned in Section 4.3 can be the selection of the combined parameters δf and x_{OV}, which define the total spectrogram size. High frequency resolutions can quickly create an interpretation problem: For example with an IF span of $3\,\mathrm{GHz}$ a resolution of $\delta f = 10\,\mathrm{kHz}$ creates $3 \cdot 10^5$ frequency bins in each spectrum, which slows down visualization and analysis considerably. Generally it is good practice to balance the two dimensions, i.e. to increase the overlap factor when very high frequency resolutions are used, because otherwise "trivial" spectrograms containing only a few spectra can occur.

4.7 Spectrogram post-processing and time extension

4.7.1 Frequency extraction

A spectrogram conveys a wealth of information about time-dependent spectral behavior to the user. Still, the signal development is only contained in an implicit way, as the data consists of intensity values in a time-frequency matrix. For any comparison of signal developments which extend beyond an optical inspection of the spectrogram, further processing is necessary.
The extraction of 1-D f-vs.-t-information is one basic technique to enable simple comparisons with other frequency measurements and automatic analysis. It was implemented for the PSA in [Sch14b], following classical object detection solutions, using image processing techniques. This retains intuitiveness in the process and allows easy assessment of the result.

Three subsequent analysis steps are employed:

- As a first step, a base noise cut through adaptive filtering is applied. This fundamentally influences the subsequent steps, since it defines the number and integrity of isolated pixel groups under investigation.

- Pixel group size based filtering using binary indicator masks allows the parametrized removal of remaining "salt-and-pepper-noise" from the previous cut. An appropriate size of the pixel groups can only be decided on the basis of time-frequency parameters of the spectrogram and the desired result. In a spectrogram with $\delta f = 1\,\mathrm{MHz}$, $x_{\mathrm{OV}} = 0.5$, for example, a normal-width signal of 20 µs duration forms a group of approximately 50–120 pixels.

- The remaining pixel groups are considered to be relevant, so a segment linking algorithm is applied to combine them into traces. Since the basic signal is a sinusoid with a discrete time-dependent frequency, the expected result are curve-type contours which are aligned more or less along the time axis. For the linking process, object orientations and centroids are evaluated which define the direction and cone angle in which the algorithm searches for more segments of the same curve.

The obtained data allow efficient comparison of different frequency or spectrum measurements, since only in traced data is an explicit information on frequency present. Of course also the total amount of data is drastically reduced, as noise is no longer contained.

Such post-processing could form a basis for modulation investigation, and for automated processing of spectrograms with the goal of establishing a database of measured spectral behavior, enabling a summarized analysis of many measurements. In this work, however, the frequency tracking was only employed as a prerequisite for the undersampling technique described in the following.

4.7.2 Operation with undersampling

Background

One profound limitation of the PSA system as presented in Section 4.1 lies in the maximum acquisition length, which is limited to 280 MS. This yields $t_a \leq 28\,\mathrm{ms}$ even when using only one receiver channel, sacrificing the 2-channel unambiguity.
A way to cover longer pulse durations could be provided through repetitive acquisition of short spectra with intermediate download from the oscilloscope, e.g. using frames of 1–5 MS. This would yield data-transfer time limited repetition rates in the range 300–1000 ms, essentially emulating a triggered spectrum analyzer. Some interesting processes, however, take place within some hundreds of milliseconds up to a few seconds, especially the transition from short- to long pulse operation, as described in Section 2.4.3.
When only the tuning and the modulation of a single signal is targeted, a drastic step can be made: By limiting the acquisition to only one channel at a reduced sampling rate $f_r' = f_r/k$, the acquisition time will increase in proportion to k. Thus, for a reduction from 10 GS/s to 100 MS/s, with $n = 100$ the acquisition time is enhanced to $kt_a = 2.8\,\mathrm{s}$.

Aliasing

This of course also reduces the Nyquist frequency by the same factor, and since the hardware-implemented low-pass at the oscilloscope front-end remains unchanged, the acquisition is now unprotected against aliasing.[16] Therefore it is important to know at which frequency f_0 an aliased signal will appear after the sampling process. The basic relationship can be found in (A.8), but is now illustrated in Figure 4.7 for a more detailed investigation.

[16] When reducing the sampling rate f_r while maintaining constant frequency resolution δf, as a secondary effect the dynamic range is also reduced. This is due to a loss of processing gain, and obvious from (A.15), considering that the number of samples N in each DFT is $N = f_r/\delta f$.

For an incoming signal with the frequency $f_{\mathrm{i}}(t)$, the detected signal $f_{\mathrm{d}}(t)$ has a different frequency if aliasing occurs. The different Nyquist zones (compare Section A.1.1) are separated by the Nyquist frequency f_{N}. A transition between two Nyquist zones in $f_{\mathrm{i}}(t)$ causes $f_{\mathrm{d}}(t) = 0$ or $f_{\mathrm{d}}(t) = f_{\mathrm{N}}$. In between the transitions, the detected signal contains the full original information, but shifted and possibly mirrored.
It suffices to know the Nyquist zone number N for each interval along the time axis to fully restore $f_{\mathrm{i}}(t)$ from $f_{\mathrm{d}}(t)$, using the properties of the alias relationship given in (A.8). With the Nyquist zone number definition $N = \lceil \frac{f_{\mathrm{i}}}{f_{\mathrm{N}}} \rceil$, the following rules between the two signals can be established from this relationship:

$$f_{\mathrm{d}} =(-1)^N \left(2\lfloor\frac{N}{2}\rfloor f_{\mathrm{N}} - f_{\mathrm{i}}\right) \tag{4.11}$$

$$f_{\mathrm{i}} =2\lfloor\frac{N}{2}\rfloor f_{\mathrm{N}} - (-1)^N f_{\mathrm{d}} \quad . \tag{4.12}$$

For a continuously tuning input frequency $f_{\mathrm{i}}(t)$, it suffices to know N at only one time position. At this reference point, the real frequency can be directly calculated from (4.12). The rest of the signal is then defined, because the Nyquist zones are necessarily adjacent to each other.
This also explains the necessity of the tracing approach discussed in Section 4.7, as the described operation cannot be "blindly" applied to spectrograms.

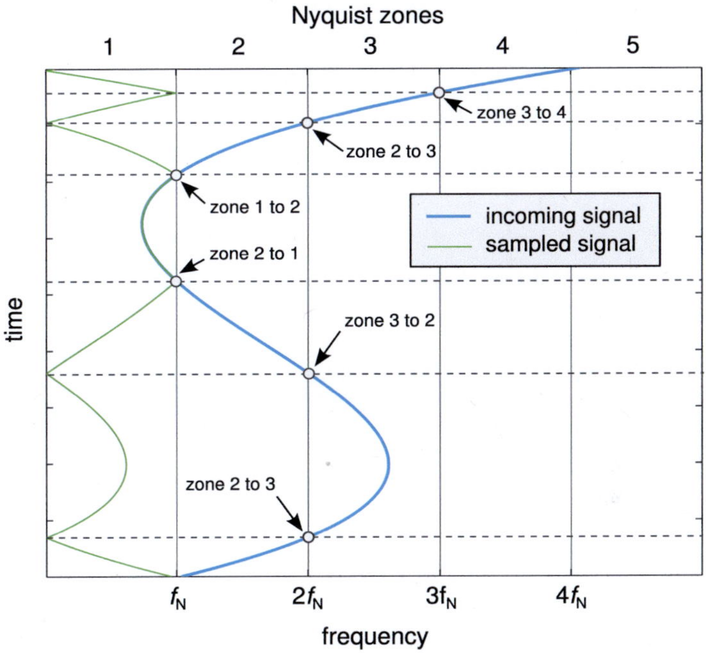

Figure 4.7: Illustration of the aliasing effect in the detected signal f_d after undersampling the incoming signal f_i (adapted from [Sch14b])

5 Primary measurement results

The application of the PSA system described in the previous Chapter to gyrotron monitoring results in a wealth of documented spectral phenomena, a selection of which is presented in the following. After a look at the typical cavity mode appearance under normal conditions, the examples move to cases of unstable and pathological behavior. It should be noted that the latter took normally place at extreme operating points which were intentionally chosen for the investigation of these phenomena.

Most of the data presented were collected during factory acceptance tests (FAT) of various W7-X series 140 GHz gyrotrons. However, some results from other gyrotrons are also included; namely these are the KIT frequency step-tunable D-band gyrotron, the KIT 170 GHz 2 MW coaxial cavity gyrotron and a 118 GHz gyrotron operating in the ECRH system of TCV[1].

5.1 Nominal cavity mode appearance

In the following, multiple examples of typical operation, taken from W7-X gyrotron experiments at KIT, are presented. Examples ranging from short-pulse operation of a few hundred microseconds pulse length, the transition to long pulse operation and the behavior during half-hour pulses are included.

[1] TCV, *Tokamak à Configuration Variable*, is a plasma fusion experiment located at CRPP (Centre de Recherches en Physique des Plasmas) of EPFL (École Polytechnique Fédérale de Lausanne) in Switzerland, operating a multi-megawatt ECRH and ECCD system at 82 GHz and 118 GHz.

5.1.1 Typical short-pulse operation

During gyrotron testing, the external parameters V_{cath} and I_{b} are ramped to the desired values by the high-voltage power supply within a few hundred microseconds. Since the electron cavity transit times are of the order of nanoseconds (compare Section 3.2.2), the applied voltage slew rate of $80\,\mathrm{kV}/0.3\,\mathrm{ms} \approx 240\,\mathrm{MV}/\mathrm{ms}$ is a very gently inclined ramp in terms of the gyrotron interaction. Because of this, the cavity mode starts as soon as the necessary starting V_{cath}-I_{b}-combination is reached, and continuously varies with the parameter ramp.
In Figure 5.1, this process can be observed in the spectrogram. When the high-voltage flat-top is reached, the cavity mode $\mathrm{TE}_{28,8}^{-}$ assumes stationary and clean operation at $140.235\,\mathrm{GHz}$.
The fact that the mode is tuned from higher frequencies is explained by means of the relativistic cyclotron frequency: With rising $V_{\mathrm{cath}}(t)$, the relativistic factor γ increases, leading to a reduction of the relativistic electron cyclotron frequency $\Omega_{\mathrm{r}} = \Omega_0/\gamma$.

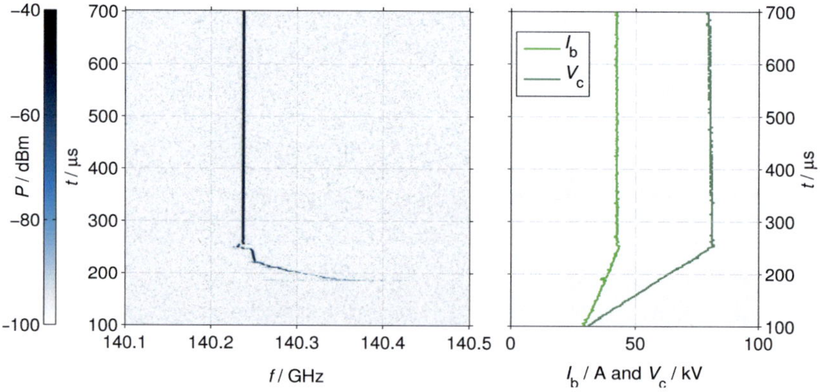

Figure 5.1: Typical short pulse behavior (W7-X gyrotron SN7) at nominal parameters; the start of the cavity mode along the V_{cath}-I_{b}-ramp and the settling at flat-top can be directly observed ($\delta f = 1\,\mathrm{MHz}$, $x_{\mathrm{OV}} = 0.3$)

5.1.2 Start of long pulse operation

SP measurements of a few milliseconds yield very homogeneous spectrograms as presented in the previous section, but cannot extend far beyond $t_p = 10\,\mathrm{ms}$ because of memory limitations. The undersampling technique introduced in Section 4.7.2 provides longer time captions while maintaining spectral resolution, allowing the investigation of phenomena with longer duration. However, the loss of unambiguity due to the undersampling process as described previously must be considered.

In Figure 5.2, the continuously recorded first two seconds of a $500\,\mathrm{kW}$ pulse with the W7-X gyrotron SN2i are displayed both on linear and logarithmic time axes. The mentioned undersampling and frequency tracing technique presented in Section 4.7 yields one-dimensional data instead of a spectrogram, so the final result as displayed here is similar to a measurement with a Modulation Domain Analyzer (compare Section 1.3).

The diagram contains the frequency tuning curve which is typical for long pulse operation. Interestingly, the gyrotron does not tune continuously, as would be expected from the consideration of the tuning effects described in Section 2.4.3, but in discrete jumps of mostly $1–5\,\mathrm{MHz}$ width. It must be stressed that this is not connected with simple measurement uncertainty or measurement system instabilities, but constitutes physical behavior of the measured signal.

Histogram analysis of the jump height distribution did not determine a specific characteristic jump height. Most jumps have heights of $1\,\mathrm{MHz}$ or less, while the larger jump heights occur equally distributed up to $10\,\mathrm{MHz}$.

One possible explanation for this behavior is the long-line effect discussed in Section 2.4.4. The frequency relations attributed to this effect as given in (2.26) however do not match, as the measured jumps are smaller by at least a factor of four than the lowest characteristic frequency which can be expected from an external long-line effect.

It is also conceivable that the resonator itself does not expand entirely smoothly, but exhibits intervals of continuous tuning between which the built up internal stresses cause slipping of grain boundaries or abrupt changes of micro-rupture patterns in the material. To give a crude estimation, for producing the jumps of 5 MHz, only slips of

$$\frac{r_{c,2}}{r_{c,1}} \approx \frac{f_1}{f_2} \quad \rightarrow \quad \Delta r_c \approx \left(\frac{f_1}{f_2} - 1\right) r_{c,1} \approx 700\,\mathrm{nm} \tag{5.1}$$

are necessary. This is not unrealistic, especially considering that the cavity of W7-X gyrotrons is fabricated of GlidCop[2]: Its locally non-uniform ductility, caused by the alumina embedded in its crystal structure, could be the principal source of the observed behavior [Sch14a].[3]

[2]Glidcop [scm73] is a copper alloy strengthened with alumina oxide, improving its yield strength and other mechanical parameters especially for high temperature operation and in high radiation environments.

[3]This matches experiments with a $TE_{10,4}$ gyrotron in [DBI+96], where after operation with very high wall loads, plastic deformation of the gyrotron resonator was found.

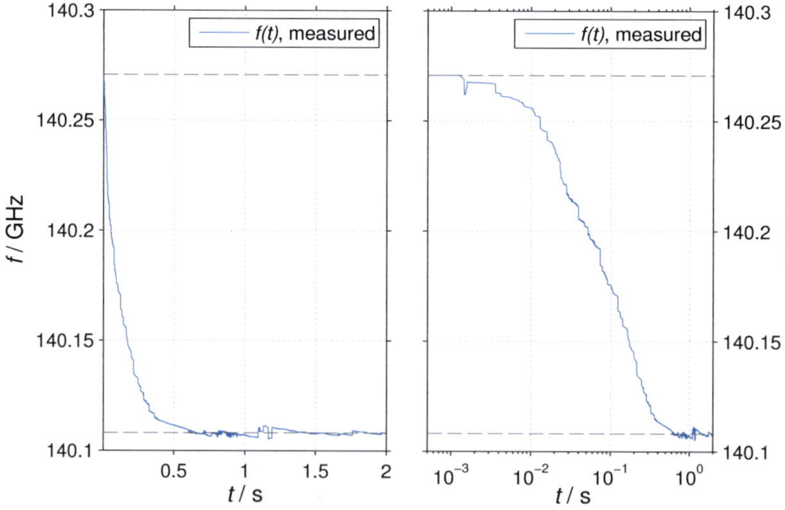

Figure 5.2: Frequency tuning at the start of long pulses, measured with deliberate undersampling (compare Section 4.7.2)

5.1.3 Long-pulse operation monitoring

During FATs, W7-X gyrotrons are operated at 500 kW for up to 30 minutes, and at up to a megawatt for 3 minutes. Such operation cannot be monitored continuously with the PSA, only segmented samples can be taken (compare Sections 4.1 and 4.7).

In Figure 5.3, eight selected samples from the same 30 min, 500 kW pulse of the W7-X gyrotron SN2i are plotted. Again a reduced sampling rate was employed; however, the measurement could be done in the first Nyquist zone because of the stationary operation. Therefore, the presented diagrams are spectrograms instead of f-vs.-t-curves as used in Figure 5.2.

Even though the measurements are taken deep within the pulse time which is expected to be stationary, the observed behavior is not continuous. Comparing the different pulse index times we find various instabilities: Staggering, discrete jumping and continuous tuning up and down.

This indicates a bi- or multi-stability between closely related stable operation conditions of the $TE_{28,8}^{-}$ mode. As the trigger for this, several effects, which are likely to act in combination, can be considered:

- **Long-line effect**
  Here this effect can be expected to be stronger than in shorter pulses, because the load and other components heat up, adding slight alterations to the beam path length through thermal deformation.
  The jump width of the range 1–10 MHz again cannot be directly associated with the transmission line length: While the total beam path from resonator end to load end at KIT W7-X gyrotron experiments is around 7 m, the jump frequencies correspond to distances in the range of about 15–150 m, following (2.26). Multi-reflections along the line are of course thinkable, but then a periodic pattern corresponding to the various round-trip times associated with the relevant reflection position in the observed jump frequencies should be observable.

- **Bistable resonator tuning**
  For the pulse time indices $t_{\mathrm{p}}/\mathrm{min} = [6, 9, 18, 22]$, a bi-stable operation could be the case. Consider the resonator at a quasi stationary operating point with $r_{\mathrm{c}} = r_{\mathrm{c},1}$ to expand slightly due to the thermal load from the oscillation, leading to a wall radius $r_{\mathrm{c},2}$ which favors a slightly different and less efficient operation of the nominal mode; changes in the percentage power range would be more than sufficient. Consequently, less power is produced and deposited in the wall, and the resonator slightly contracts back towards $r_{\mathrm{c},1}$, resulting in favorable conditions for the initial operation.
  Such a mechanism could also explain the fact that the slopes of the f-vs.-t ramps in the mentioned measurements are opposed in between jumps.

- **Acoustic vibration**
  All power guiding components and especially the load are flown through by hundreds of liters of cooling water per minute at pressures of multiple bar. This and the adjacent pumps could easily cause parts of the system to vibrate with up to multiple kHz, which could also cause slight instabilities like the ones presented.

To decide which of the listed mechanisms is the cause for the frequency jumping is hard; an important point made here, however, is that the effect cannot be caused by the measurement system.

Apart from the discrete sub-second effects above, a small but measurable tuning of the average frequency over the whole duration of the pulse can be observed. From the measurements in Figure 5.3, the shift rate can be estimated to be approximately $1.5\,\mathrm{MHz}/\mathrm{min}$. An explanation is found in the cooling system: Due to its finite cooling power, the inlet coolant temperature slowly rises during the pulse, allowing the temperature of all cooled components to rise slowly and linearly. This again tunes the resonator continuously.

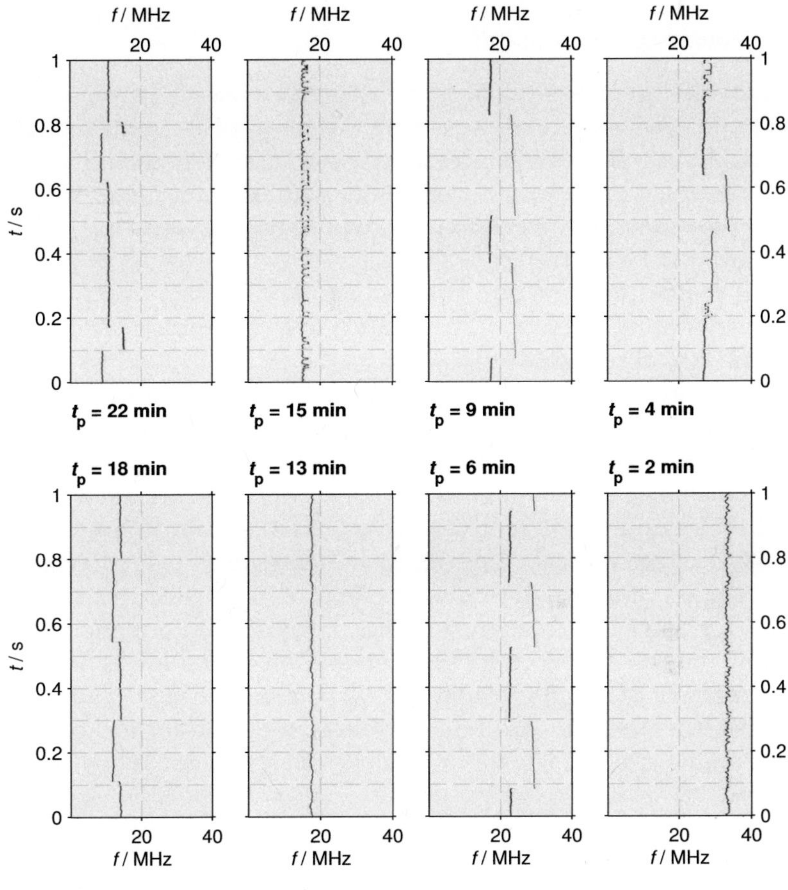

Figure 5.3: Periodic long-term monitoring during LP operation. All sample spectrograms were obtained at the given pulse times t_p during the same pulse with an aliasing-free span of 100 MHz, and processed with $\delta f = 50\,\mathrm{kHz}$, $x_\mathrm{OV} = 0$. The frequency axis indicators are relative to the absolute frequency 140.05 GHz.

5.2 Cavity mode instabilities

5.2.1 Power supply voltage ripple

While the KIT gyrotron test-stand incorporates a tetrode power supply with a specified voltage ripple of 0.3 kV peak-to-peak, many modern test-stands, as well as the ITER facility, use solid-state stacked-thyristor power supplies. This introduces higher and more pronounced voltage ripples above 1 kV peak-to-peak, whose impact on the operating frequency can be measured.
During a measurement campaign at the TCV facility, cavity mode frequency traces as displayed in Figure 5.4 were recorded at a 2 s CW gyrotron operating at 118 GHz.
On the cavity mode frequency, a distinct 10 MHz ripple with a periodicity of 200 μs is visible. The anode voltage as recorded with a fast oscilloscope is displayed in the additional parameter plot. Note that the visible noise on the measured trace is not the actual noise of the HVPS, since unfavorable combinations of high voltage divider ratios and oscilloscope vertical ranges easily induce strong apparent noise.[4]
The averaged version of the same data shows a more direct relationship between voltage and frequency ripple. A qualified information however can only be derived from the spectrogram on the left, as here the direct effect of the ripple is observed. Of course for heating purposes, such ripples are of no concern. However they can further complicate certain diagnostic situations, where clean frequency knowledge is required.
The mentioned gyrotron also exhibited parasitic oscillations, which peculiarly were found at frequencies above the cavity mode's, at 119.8 GHz and 120.6 GHz, mutually excluding each other. Interestingly, their frequencies were practically unaffected by the instability of the external voltages, which is hard to explain. One possibility might be a very high Q of the oscillations [Alb12].

[4] For the application of plasma heating, such effects are not of primary concern. They, however, could possibly cause disturbances in other diagnostic systems, and of course complicate the investigation of the gyrotron itself. Also, the comparison with simulations becomes more complicated.

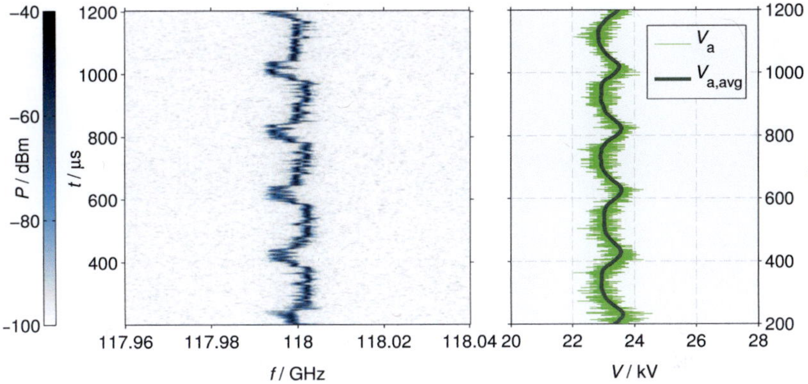

Figure 5.4: Excerpt from measurement during 1 s-long pulse operation of a TCV 118 GHz gyrotron ($\delta f = 500\,\mathrm{kHz}$, $x_{\mathrm{OV}} = 0.2$)

5.2.2 Short-time stability

During an experiment on asymmetry between the mechanical axis of the gyrotron and the magnetic field with the KIT frequency step-tunable D-band gyrotron [KDK+01], the mode stability and the rise of modulation sidebands was monitored with the PSA system.

The described operation took place with the $\mathrm{TE}^-_{22,8}$ cavity mode at $V_{\mathrm{cath}} = 88.3$–$90.3\,\mathrm{kV}$ and $I_{\mathrm{b}} = 41\,\mathrm{A}$. Using the gun coils, the pitch factor α was increased. While no parasitic oscillations were detected, with rising alpha the short-time stability of the cavity mode decreased, while the power remained almost constant. This is illustrated in the figure series 5.5, 5.6; here the end of the power supply ramp-up (compare Fig 5.1) is indicated at $t_{\mathrm{p}} = 280\,\mathrm{\mu s}$.

Depending on the STFT resolution, this appears as a mere line widening, or a "staggering" behavior. The modulation sidebands have different frequencies; however, basic common principles must apply, because the same frequencies keep repeating in different combinations. Note that this staggering effect is an incoherent property, and does not exhibit the periodicity of the ripple-induced instability in Section 5.2.1.

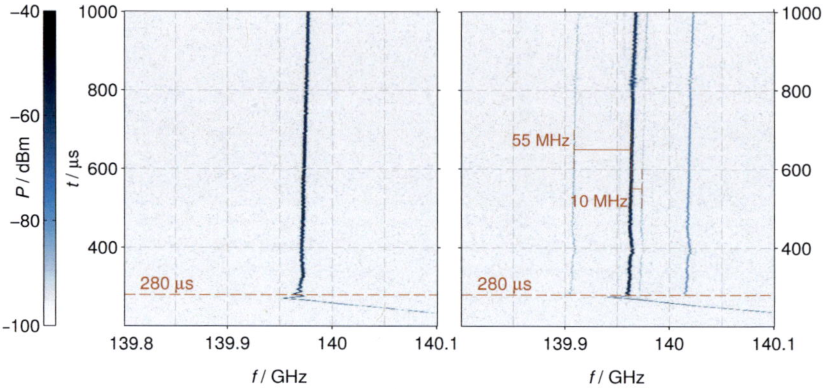

Figure 5.5: Spectrograms ($\delta f = 100\,\mathrm{kHz}$, $x_{\mathrm{OV}} = 0.5$) of instabilities in the KIT frequency step-tunable D-band gyrotron with
left: $\alpha = 1.39$ at $P = 770\,\mathrm{kW}$
right: $\alpha = 1.44$ at $P = 780\,\mathrm{kW}$

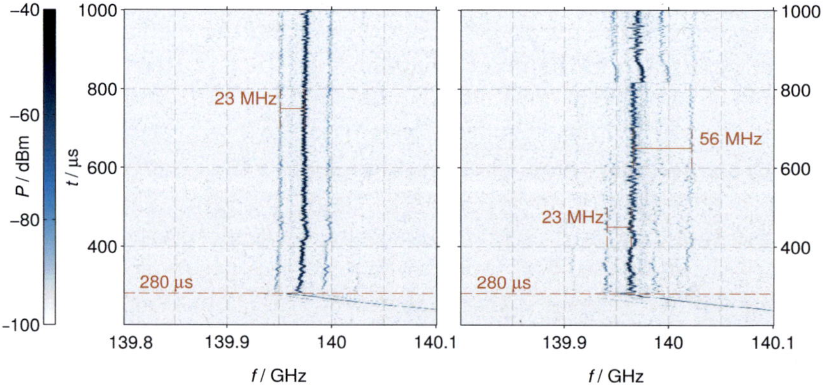

Figure 5.6: Spectrograms ($\delta f = 100\,\mathrm{kHz}$, $x_{\mathrm{OV}} = 0.5$) of frequency-staggering of the KIT frequency step-tunable D-band gyrotron accompanied by increasing LF modulation with
left: pitch factor $\alpha = 1.51$ at $P = 820\,\mathrm{kW}$
right: pitch factor $\alpha = 1.62$ at $P = 840\,\mathrm{kW}$

The effect shows some similarity to "periodic noise" caused by trapped ions in the electron beam range, as investigated e.g. in [MAC+03] for TWTs and gyro-klystrons, where phase and slight power instabilities in the 10–1300 kHz range are observed. However, in the measurements presented here, the instability is visible immediately after the start of the pulse, without much change during a multi-millisecond pulse. An effect based on the ionic charge should be related to the characteristic time constants of ion generation, which are much longer than the pulse durations considered here.[5]

5.3 Cavity mode competition and switching

5.3.1 Cavity mode switch in a W7-X gyrotron

When operated at the edge of stability, a cavity mode can start to compete with a neighboring mode. This can lead to reduced power, or even to a cavity mode switch (compare Section 2.4.1). Such effects are mostly treated with simulations; however, with the PSA system presented in this work, they can be investigated experimentally.
In Figure 5.7, excerpts of a spectrogram captured during operation of W7-X series tube SN5 at the edge of stability of the nominal cavity mode $\mathrm{TE}_{28,8}$ are depicted. After a normal start-up until $t_\mathrm{p} = 240\,\mu\mathrm{s}$, immediately unstable behavior is observed: There is a very fast and short start-up attempt of the neighboring mode, visible by the extremely short and broad-band signature in the 137.5 GHz range in combination with a spike in the nominal mode at $t_\mathrm{p} = 245\,\mu\mathrm{s}$. Also the $\mathrm{TE}_{28,8}$ mode has weak sidebands and exhibits incoherent instability (compare with Fig. 5.1). The mode switch happens at $t_\mathrm{p} = 629\,\mu\mathrm{s}$, where the $\mathrm{TE}^-_{28,8}$ mode ends abruptly and is replaced by a mode at 137.51 GHz, very probably the $\mathrm{TE}_{27,8}$ mode, which continues until the pulse end at $t_\mathrm{p} = 1066\,\mu\mathrm{s}$. This behavior is the most simple case of mode competition: Within a very short time interval ($\Delta t < 1\,\mu\mathrm{s}$) the operating mode is entirely lost, and the competitor establishes stationary operation.

[5] See Section 2.4.3 and especially Chapter 7 on this subject.

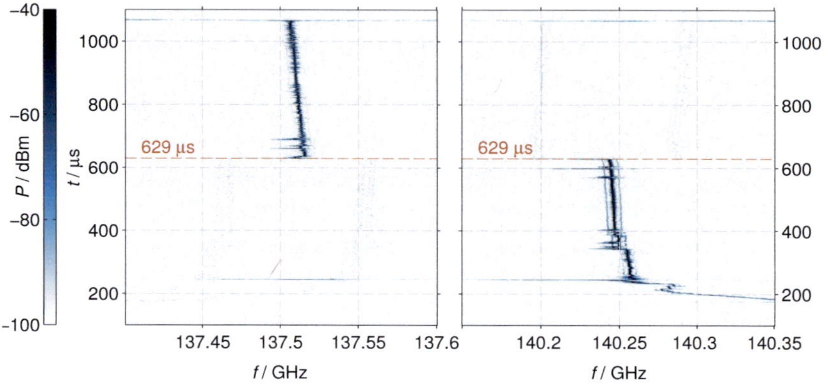

Figure 5.7: Spectrograms ($\delta f = 400\,\mathrm{kHz}$, $x_{\mathrm{OV}} = 0.5$) showing an abrupt cavity mode switch from the nominal cavity mode $\mathrm{TE}^{-}_{28,8}$ to the azimuthal neighbor $\mathrm{TE}_{27,8}$ in W7-X gyrotron SN5; the V_{cath}-I_{b} flat-top was reached at $t_{\mathrm{p}} = 240\,\mathrm{\mu s}$

5.3.2 Parallel cavity modes and LF modulation in a W7-X gyrotron

During experiments at KIT, in various gyrotrons low-frequency modulation (LFM) effects could be observed. A very good example is presented in Figure 5.8, where originally mode-switching at the edge of stability of the nominal cavity mode in W7-X gyrotron SN5 was under investigation.

The nominal $\mathrm{TE}^{-}_{28,8}$ cavity mode rises with the power supply ramp, which reaches flat-top at $t_{\mathrm{p}} = 220\,\mathrm{\mu s}$. Immediately, strong modulation patterns with $f_{\mathrm{M}} = 4\,\mathrm{MHz}$ line spacing appear. Also, a second strong oscillation around $137.5\,\mathrm{GHz}$ rises, exhibiting similar modulation. From simulations, this mode is most likely the co-rotating lower azimuthal neighbor mode $\mathrm{TE}^{-}_{27,8}$. Both modes are operating in parallel for $1\,\mathrm{ms}$, after which the LFM ends and parallel operation with a presumably dominant $\mathrm{TE}^{-}_{27,8}$ mode follows (compare Section 2.4.1 and [ZSST06]). Interestingly, multiple clearly defined LFM segments

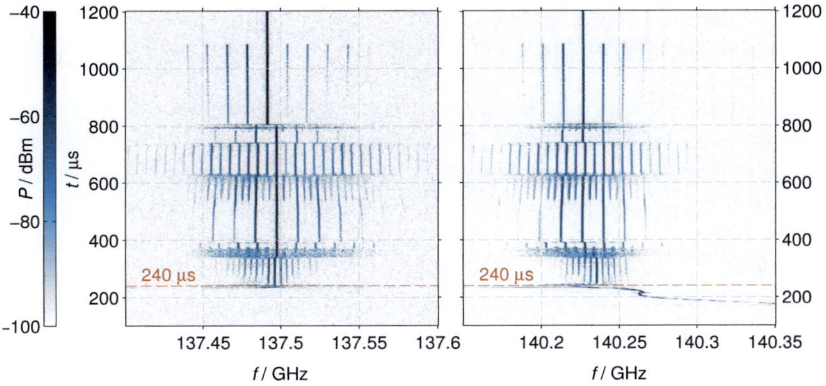

Figure 5.8: Multi-mode operation with strong LFM ($\delta f = 300\,\mathrm{kHz}$, $x_{\mathrm{OV}} = 0.5$)

can be distinguished. In these intervals, the sideband spacing δf_M changes discretely between $4, 6$ and $12\,\mathrm{MHz}$, alternating between the values. With the modulation frequency changes, at some points also the "carrier frequency" jumps by a few MHz.

Another interesting modulation feature is the phenomenon of "period multiplication": Moving from $t_{\mathrm{p},1} = 400\,\mathrm{\mu s}$ with $f_{\mathrm{M},1} = 12\,\mathrm{MHz}$ in the positive time direction, period tripling can be observed around $t_{\mathrm{p},2} = 540\,\mathrm{\mu s}$ with $f_{\mathrm{M},2} \approx f_{\mathrm{M},1}/3$. This is followed by period doubling with respect to the frequency $f_{\mathrm{M},1}$ at $t_{\mathrm{p},3} = 635\,\mathrm{\mu s}$. After this, $f_{\mathrm{M},1}$ is restored at $t_{\mathrm{p},4} = 740\,\mathrm{\mu s}$. Similar behavior has also been described in [AAA+12] for a THz-gyrotron operating with $0.2\,\mathrm{kW}$ in the $\mathrm{TE}_{7,2}$ mode at $260.5\,\mathrm{GHz}$; above example, however, is the first detailed documentation of this effect in a high-power millimeter-wave gyrotron.

The measurement of strong cavity mode LF oscillations as well as parallel cavity mode operation has also a certain secondary importance, as such behavior is found occasionally in self-consistent simulations, and often raises suspicions of numerical issues. It must, however, be noted that because of the instability of the operating point, the pulse-to-pulse reproducibility of this behavior was very low.

5.3.3 Parallel cavity modes and LF modulation in the 2 MW coaxial gyrotron

Competitive cavity mode instabilities were also observed in experiments with the KIT 2 MW, 170 GHz coaxial gyrotron [RPK+10], operating at the $\mathrm{TE}^{-}_{34,19}$ cavity mode. The basic situation is illustrated in Figure 5.9, and is relatively similar to the one in the cylindrical-cavity W7-X gyrotron at the edge of stability: A lower azimuthal neighboring mode $\mathrm{TE}_{33,19}$ is present, and is operating in parallel to the nominal cavity mode. The described operation was monitored at $V_{\mathrm{cath}} = 82.4\,\mathrm{kV}$ with a beam current $I_{\mathrm{b}} = 63.1\,\mathrm{A}$.

In the RF spectrogram, the line width of the signal is higher than in other measurements, around 3 MHz to 5 MHz at a resolution bandwidth of 1 MHz. While the nominal mode is active around 170.04 GHz, the neighbor competitor oscillates in intervals just below a frequency of 168 GHz.

The possibility of a posteriori Fourier analysis is applied here, and shown in Figure 5.10. A zoom on the cavity mode (right spectrogram in Fig. 5.9) with an enhanced resolution bandwidth of 50 kHz and high overlap factor shows that the cavity mode is modulated with a low bandwidth. Three selected spectra from different pulse time indices are shown in the left part of Figure 5.10.

While at $t_{\mathrm{p}} = 500\,\mu\mathrm{s}$ a quite typical linewidth of $\sim 1\,\mathrm{MHz}$ is found, at $t_{\mathrm{p}} = 870\,\mu\mathrm{s}$ close sidebands at a distance of roughly 200 kHz can be observed. They resemble the typical Bessel-style spectrum of analog frequency modulation with medium modulation index. The spectrum at $t_{\mathrm{p}} = 1667\,\mu\mathrm{s}$ also fits into this type, however for the case of a higher modulation index.

It should be noted that the LF pickup loop antenna in the gun region detected low-frequency oscillations, however, only in bursts at 113 MHz and around 600 MHz. The modulation frequencies from the RF measurement were not found in the LF signal, and no modulation corresponding to the LF signals could be found. This underlines the complexity of low-frequency modulation phenomena, and the care which is necessary during their investigation.

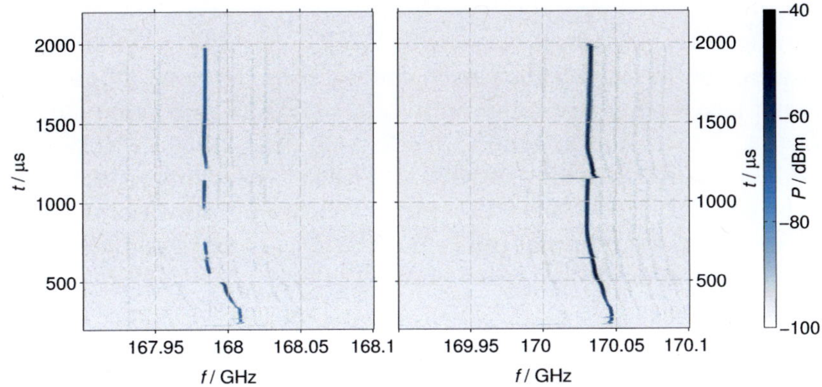

Figure 5.9: Overview of 1.7 ms pulse by the 170 GHz KIT coaxial cavity gyrotron with $\delta f = 1\,\mathrm{MHz}$, $x_{\mathrm{OV}} = 0.3$

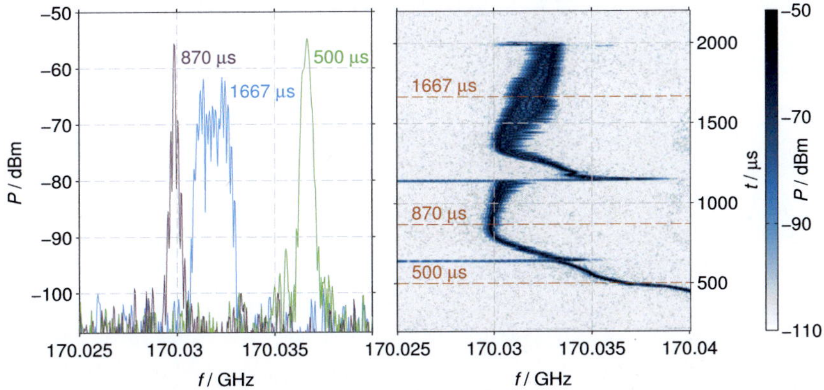

Figure 5.10: Right: reprocessed detail of Figure 5.9 with $\delta f = 50\,\mathrm{kHz}$, $x_{\mathrm{OV}} = 0.95$
Left: separate spectrum excerpts (t_{p} indices as indicated)

5.4 Spectral distortions during a window arc

During early SP experiments with W7-X tube SN5, a window arc occurred at the edge of the nominal mode's stability range. The event took place during operation with $V_{\mathrm{cath}} = 78.9\,\mathrm{kV}$, the beam current $I_{\mathrm{b}} = 41.3\,\mathrm{A}$, a cavity magnetic flux density of $B_{\mathrm{z,cav}} = 5.56\,\mathrm{T}$ and an electron beam radius of $r_{\mathrm{b}} = 10.25\,\mathrm{mm}$. The pitch factor was calculated to be equal to 1.21. During first pulses at this operating point, stable operation of the nominal cavity mode $\mathrm{TE}^{-}_{28,8}$ at 140.25 GHz with an output power of 715 kW was measured.[6]

5.4.1 Observed RF behavior

In Figure 5.11, an excerpt of the RF spectrogram during the described event is given, along with the logged values of the cathode voltage V_{cath} and the beam current I_{b} on the common vertical time axis.
In the time interval $180 < t_{\mathrm{p}}/\mu\mathrm{s} < 260$, the power and frequency dependence of the $\mathrm{TE}^{-}_{28,8}$ mode on the external parameters V_{cath} and I_{b} can be observed. From $t_{\mathrm{p}} = 260\,\mu\mathrm{s}$, the power supply having reached flat-top with 78.9 kV and 41.3 A, stable operation of the nominal cavity mode $\mathrm{TE}^{-}_{28,8}$ at 140.26 GHz can be seen. At $t_{\mathrm{p}} = 320\,\mu\mathrm{s}$, another oscillation around 137.45 GHz rises quickly and competes with the nominal mode. Both modes exhibit strong modulation, visible in the spectrogram as a 16 MHz-sideband "lattice pattern". This activity is abruptly ended with the loss of the $\mathrm{TE}^{-}_{28,8}$-mode at $t_{\mathrm{p}} = 355\,\mu\mathrm{s}$, which is followed by stable unmodulated single-mode operation of the competitor with $f = 137.52\,\mathrm{GHz}$ until the pulse ends at $t_{\mathrm{p}} = 370\,\mu\mathrm{s}$.
The shut-down was caused by a window arc alert. Taking into account the measured interlock latency of 50–70 µs, the arc must have started in the time interval $305 < t_{\mathrm{p}}/\mu\mathrm{s} < 325$. This coincides with the onset of perturbed RF operation, but it is in the first place hard to determine whether the arc was caused by the mode competition or vice versa.

[6] This operating point is part of a parameter study, and therefore does not constitute optimal power generation efficiency. With the correction means derived in Section 3.1.3, the available power at the cavity exit can be estimated to be $720 \lesssim P_{\mathrm{dif}}/\mathrm{kW} \lesssim 830$.

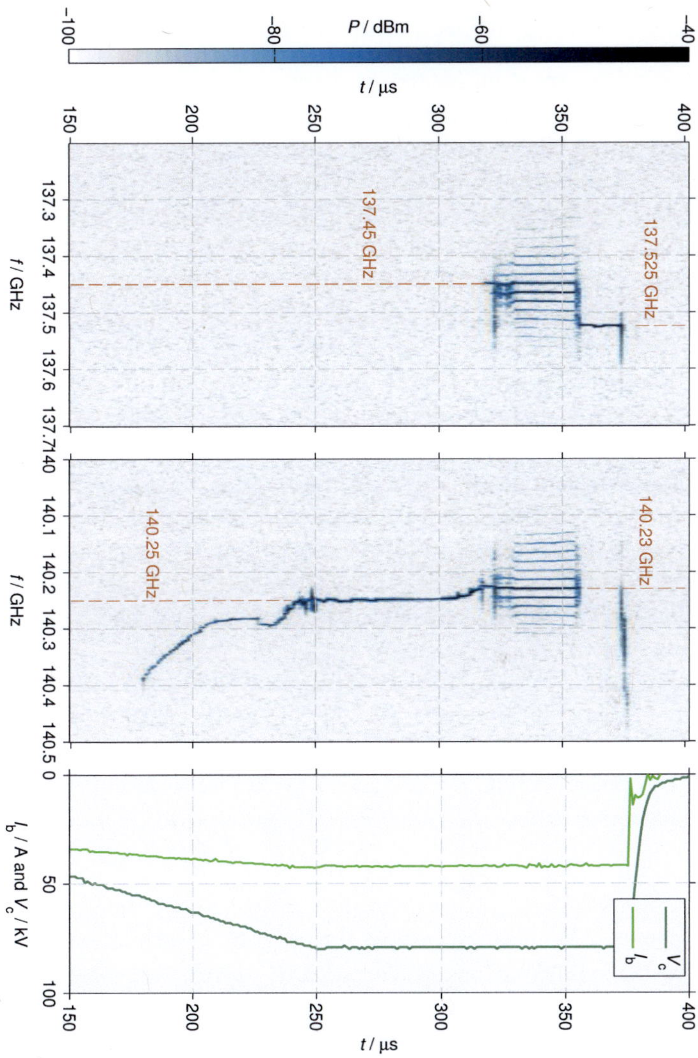

Figure 5.11: Measured spectrogram of a gyrotron pulse during window arcing; the start-up phase, a stable oscillation interval of the $TE^{-}_{28,8}$ mode at 140.27 GHz and destabilization with mode competition are directly observable

5.4.2 Discussion of RF behavior during window arc

An RF window arc is an electrical discharge following a microwave-induced breakdown at the gyrotron's dielectric window. The arc is being fed power through the microwave beam and can easily break the RF window through sudden thermo-mechanical stress, effectively destroying the gyrotron [DZLM03]. For quasi-optical transmission lines, this is typically an air breakdown on the atmospheric side of the dielectric window. Not many properties of RF arcs at millimeter-wave frequencies are known; the few but impressive systematic investigations available have established that complex effects such as two-dimensional plasma filament array formation and motion along the microwave beam axis take place [HCM+08, HCM+09].
Of the RF power P_0 available at the window, a part P_{pass} will pass through the arc and reach the calorimeter. Since the arc is sustained by the RF power, it will absorb a part P_{arc} and act as an additional power sink between the gyrotron and the load. Parameter sweep experiments in fundamental-mode waveguides at lower frequencies indicate that approximately 50–80% of the RF power can be absorbed by an RF arc [DT03]; however, these values are still vague. Assuming that the arc is forming a planar plasma filament array parallel to the phase fronts of the RF beam as described in [HCM+09], in a first-order approximation its reflective properties can be expressed by a complex reflection factor $\underline{r}$, defining the reflected power P_{ref}. A power balance maintaining the convenient properties of a reflection factor can thus be formulated as

$$P_0 = P_{\text{ref}} + P_{\text{arc}} + P_{\text{pass}} \tag{5.2}$$

with $P_{\text{ref}} = |\underline{r}|^2 P_0$. Unfortunately, the ballistic calorimeter which was employed does only allow summarizing power measurements over the whole pulse, so that P_{pass} remains unknown. Initially the only known quantity in the above equation is P_0, so this balance cannot be used for parameter range estimations. Another interesting and unknown property is the "minimum burning power", i.e. the power which the arc needs to absorb continuously to sustain itself, which could be used to narrow the variable range in the power balance.

Concerning the gyrotron interaction inside the cavity, the mentioned power reflection is the dominant effect of the arc. The long-line effect as discussed in Section 2.4.4 is a highly probable source for the modulation pattern; however the association of the sideband distance with a geometrical location along the beam path after (2.26) fails. For a reflection from the area of the window, an expected modulation frequency of $f_M \approx 100\,\mathrm{MHz}$ is defined (compare Fig. 2.5), contrasting with the measured value of $16\,\mathrm{MHz}$. To illustrate the size of deviation, $f_M = 16\,\mathrm{MHz}$ would translate into a value of $9.38\,\mathrm{m}$ for the reflection distance[7], while the measurement was made in short-pulse operation with the ballistic calorimeter mounted directly at the window (see Section 2.4.3). The total beam path length of $1\,\mathrm{m}$ is thus considerably below the predicted value, so this interpretation does not yield any additional information.

Interestingly, this modulated operation does not continue till the pulse end shutdown, but is replaced with undistorted operation of an undesired cavity mode at $f = 137.52$ GHz, as pointed out earlier. This indicates strongly that the arc very probably ended at this point of time, before the end of the pulse itself.

5.4.3 Causal investigations

An important question which can be indirectly addressed through the advanced measurement data is the probable cause of events. Either initially the mode switch occurred for some reason and caused the arc, or an arc appeared at the window, which caused the mode switch.

The measurement yields detailed information on the spectral behavior, so the basis for simulation parameter studies attempting to reproduce the observed effects exists. Using SELFT (compare Section 3.2.2), the operating point was simulated and a two-dimensional parameter study of $\underline{r}$ was conducted in continuation runs. This emulates the latter of the

[7]This is the calculated value for for the $TE_{28,8}$ mode. However, due to the large value these quantities do not change significantly for neighboring modes, such as the $TE_{27,8}$ mode at 137.45 GHz, because the value is dominated by the speed-of-light part of the transmission line (compare Fig. 2.5).

two alternative scenarios discussed above, in that after a clean start-up with $\underline{r} = 0$, a sudden arc onset with $|\underline{r}| \neq 0$ is simulated.
It should be noted that the modeling is subject to a number of simplifications, which limit the significance of the results. The main reason is that only the cavity is simulated; hence, the modeled reflection is located at the uptaper end, while the complete internal quasi-optical transmission line is ignored. This ignores several effects:

- The distance between the interaction inside the cavity and the reflection at the window adds a considerable delay to a reflected wave returning to the cavity. This is only expressed over the phase of the reflection coefficient, instead of a true time delay.
- In the launcher, dispersion effects and mode conversion take place, causing very different conditions for different modes. As the simulated setup essentially corresponds to a gyrotron with axial outcoupling, all launcher and chirality effects are neglected.

The study was published in [SGJT13a], and can be summarized as follows: In the complex $\underline{r}$ plane, stable RF operation was found for $|\underline{r}| < 0.1$, which agrees with older load-pull investigations [Bor01]. For larger reflections, depending on the phase of the reflection factor, either stable operation with increased output power, auto-modulation, mode competition or mode switching was found. The two latter operation ranges formed a concave area in the range $0.2 \leq |\underline{r}| \leq 0.3$ in a phase sector spanning $\sim\pi/2$, and uniformly involved the $\mathrm{TE}^{+}_{27,8}$ mode operating at $137.45\,\mathrm{GHz}$.
This fits very well with the measurement, especially concerning the frequency, and indicates that the arc was the prior cause for the unstable behavior. As the found mode is of the counter-rotating $(+)$ variety, also the apparent "self-extinguishing" of the arc can be explained: Such a mode is scattered in the launcher, is thus not directed at the window in a concentrated way and of course cannot couple back into the launcher very well when reflected. This way, a switch to a counter-rotating mode deprives the arc of its power source through loss of the nominal $\mathrm{TE}^{-}_{28,8}$ mode.

6 Parasitic oscillations

This Chapter treats the phenomenon of parasitic oscillations. After introducing the necessary terminology and an overview on the considered means of investigation, the parasitic frequencies observed in W7-X gyrotrons are analyzed.
First, the measurement data gathered with the newly developed PSA system as presented in Section 4 are investigated on several levels, considering details of individual measurements as well as systematic parameter sweep ensembles and high-level data consolidations. Then the connection with analytical linear theory is made by the analysis of measurement data clusters in the Brillouin diagram. Finally, selected operating point groups from the data pool are investigated with the numerical simulation tools introduced and benchmarked in Chapter 3.

6.1 Background and classification

As already defined in Section 1.2, in this work the term "parasitic oscillations" is primarily applied to undesired gyro-interaction outside of the cavity. It is useful to distinguish between parasitic interactions which are located upstream of the cavity (*before cavity oscillations*, BCIs) and which are located downstream (*after cavity oscillations*, ACIs).
The primary effects caused by all RF parasites are listed below, while a more detailed discussion of the consequences can be found in the ensuing BCI/ACI overview sections.

6.1.1 General deleterious effects

The problems that might arise from the existence of a parasite mode depend on its power, frequency and the excitation position; the principal effects are as follows:

- **Beam parameter distortions**
  Each beam-wave interaction distorts the kinetic configuration of the electron beam. In the case of a powerful and optimized oscillation such as the one in the cavity, the primary parameters α and γ are altered by the energy extraction.
  On the other hand, even an exceptionally powerful parasitic oscillation will have a power below 50 kW, and not change the primary beam parameters significantly. In this case, the effect on the statistical distribution of the kinetic parameters is dominant. Consider e.g. a strong overbunching interaction, in which most of the extracted energy is fed back into the electron beam again: The net energy exchange could be close to zero, while with an increase of the beam parameter spread values $\delta\gamma$ and $\delta\alpha$ a severe distortion of the beam would nevertheless be present.
  As only beam distortions introduced upstream of the cavity will impair the interaction there, this effect is the one which primarily motivates the BCI/ACI classification.

- **Stray radiation effects**
  The frequency and mode of a parasitic oscillation only match the geometry at the interaction position. Thus, it is probable that the power carried by the parasitic oscillation is converted into more or less homogeneous stray radiation. Only a part of this power leaves the tube through internal loads, relief windows and the main window. Internally, this induces unforeseen thermal loads in weakly cooled areas.
  Since even a very inefficient interaction with the multi-MW electron beam creates stray radiation in the kW range, this can quickly lead to damage of components. Also, the outgassing rates from the internal surfaces strongly depend on the material temperature. The heating from the parasite stray radiation can therefore cause a degradation of the vacuum quality inside the tube shortly after the parasite onset. This indirectly triggers pulse abortions over the vacuum quality interlocks and can extend conditioning times, reducing the reliability of the gyrotron.

- **External complications**
  Effectively, a parasite appears as a spurious signal which can leave the tube and propagate into other systems. Hence, as a secondary effect, the discrete frequency created by the parasitic oscillation can lead to problems outside of the tube, depending on the application. Especially for diagnostic systems which rely on frequency counting or broadband power detection, the additional frequency lines generated by PO can cause performance degradation or lead to incorrect results.

While all of the listed effects apply to both BCIs and ACIs, a number of important differences exist, which are described in the following classification.

6.1.2 Before-cavity interactions (BCIs)

Effects

The unperturbed low velocity spread high power electron beam can easily excite a waveguide mode as soon as the geometry and the cyclotron frequency fulfill the gyro-resonance condition, which makes the area between cathode and cavity most prone to unwanted oscillations. Consequently, the most typical undesired oscillation type in high power gyrotrons are BCIs, most often referred to as beam tunnel oscillations [GDF+10, ATA+11, CBB+11, SKI+03].
Being located upstream of the cavity oscillation, the beam distortions introduced by the BCI parasite easily limit the achievable RF power of the cavity oscillation by reducing the beam quality. If a parasite starts prior to or together with the cavity mode, the beam disturbance caused by it can also modify the start-up conditions in the cavity, leading to a different final cavity mode by shifting stability boundaries with respect to the external operation parameters.
Even if the deleterious effects are minor, another practical problem is created: The alteration of the beam parameters due to the parasite are not covered in the preliminary simulations for gyroton interaction simulation (compare Section 3.2.2). This causes deviations between

simulation and experiment, which limit performance predictions and can complicate the diagnosis of other problems.
The absorbing structures in the beam tunnel (compare Section 2.3) are only meant to suppress the parasitic oscillation. If this fails, considerable thermal loadings can occur in the lossy material which can excessively heat this part of the gyrotron. For example in the past there have been instances of thermally cracked ceramic rings in the beam tunnel [GDF$^+$10, SFG$^+$10], caused by BCI parasites which were identified as high order $TE_{0,n}$ backward waves at the end of the beam tunnel.[1]
It is also conceivable that a strong backward-wave BCI parasite could heat the gyrotron gun with its RF power, increasing the beam current in a feedback relationship. However, to the knowledge of the author, this has not yet been documented for high power gyrotrons.

Investigation

The dedicated simulation of BCIs is a demanding task, as the beam tunnel volume is much larger than that of the cavity and its structure is very complex, since it is designed to prevent the rise of oscillations (see also Section 2.3). The employed design concepts, i.e. asymmetrical or pseudo-chaotic structures, lossy dielectrics, diffractive features in with dimensions in the same order of magnitude as the oscillations' wavelengths, unfortunately also create large obstacles for efficient numerical field calculations.
A thorough discussion of simulations for a copper-ceramic stack beam tunnel as it is used in W7-X gyrotrons can be found in [LIT12], where the FISHBONE-code [LVAT09] is employed. Another technique is used in [CV12], where a semi-analytical eigenmode analysis for beam tunnel structures is described. This yields field profiles, which can be investigated for their oscillation properties and associated starting currents with tools from the EURIDICE package as done in [CAV13].

[1] These parasites were removed by extending the beam tunnel design with azimuthal corrugations, breaking the cylindrical symmetry which was a prerequisite for the emergence of the problematic modes.

Next steps are full-wave treatments, which allow a more general investigation for possibly resonant modes; first investigations indicate that many relevant modes might not be of the pure TE type, but exists as a mixture of different transverse and axial eigenmodes.
One fundamental problem for simulations in the beam tunnel area is the typical mode-based treatment: For similar simulations of the cavity, the desired mode is known, and a small number of other modes can be selected for simulation based on various relevance criteria, and their relation to the nominal mode. The beam tunnel has larger dimensions and a complicated structure, making an initial mode selection quite impossible.[2] Lossy dielectrics and radius jumps cause mode conversion and diffraction, resulting easily in hybrid mode mixtures.
A solution for this problem could be true 3D PIC simulations, but the limitations already mentioned in Section 3.2.2 apply here, even more.

6.1.3 After-cavity interactions (ACIs)

While BCIs are almost a standard problem, the phenomenon of ACIs contains more diversity, and often is a subject of speculation.
In comparison to beam tunnel oscillations, there are two fundamental distinguishing features of the interaction situation for ACIs:

- Downstream of the cavity, a "spent" electron beam is encountered, whose properties have been altered on several levels by the interaction with the cavity mode: The power extraction has reduced γ and α, while the statistical properties of the beam were altered through the bunching mechanism. Investigations with codes such as ARIADNE indicate that not only a broadening, but also significant skew and asymmetry of the statistical beam parameter distribution functions occur [Pag13].
  More subtly, it can also be expected that the beam now exhibits a strong imprint of the cavity mode's azimuthal index, altering the starting conditions for independent modes according to their azimuthal similarity to the cavity mode.

[2] The work in [LIT12] in fact presents simulations with a single mode, which was selected based on experimental and theoretical conclusions, e.g. in [GDF+10].

- Direct influence on the cavity interaction can only be expected if the ACI occurs very close to the cavity, or involves a mode which is able to penetrate into the cavity as a backward wave. This makes ACIs in the first place less harmful than BCIs, which influence the electron beam upstream of the cavity.

While the beam parameter distortions from an ACI as mentioned in Section 6.1.1 cannot influence the cavity mode, they can still deteriorate the performance of the gyrotron significantly: When in CW operation a depressed collector is used, the beam electrons downstream of the interaction need a minimum kinetic energy to reach the collector wall. Electrons which are too slow are electrostatically reflected towards the gun, and can agglomerate critically in the vessel, causing for example voltage standoff-problems or LFOs. In effect, the slow electrons limit the applicable body-collector depression voltage and therefore the attainable total efficiency.
An ACI parasite bunches the electron beam during its interaction, which additionally increases the beam parameter spread beyond the value which is naturally caused by the nominal interaction in the cavity. A broadening of the electron energy profile can create fractions of electrons with lower $\beta_{\|}$ than the minimum value which was expected in the design, leading to the problems described above.
In this work, a classification of ACIs into the three different categories

- general ACI,
- static ACI and
- dynamic ACI

is proposed, which is losely based on a tentative classification made in [KAC+10a, KAC+10b]. In the following, the individual categories are defined and discussed more extensively.[3]

[3]Other publications do not explicitly distinguish between these ACI types, complicating comparison.

General ACI

This type of ACI is the one which is most similar to the BCI situation: It describes an independent gyrotron-oscillation of any arbitrary mode, extracting energy from the electron beam.[4] An additional source of RF radiation at another frequency is created, with all the possible consequences mentioned above.
It still has the peculiarity of operating downstream of the cavity, with a partly spent electron beam. This means that the interaction could depend on the bunching introduced by the cavity operating mode, e.g. causing the ACI to have the same azimuthal index as the cavity mode, or limiting the parasite to certain efficiency ranges of the cavity interaction. An important consequence of this would be that the ACI could be entirely dependent on the existence of the cavity mode.

Static or stationary ACI

The most explicitly investigated ACI effect [ZM04, CSST07, SN09] does not in fact involve a genuine gyrotron *oscillator* interaction[5]: Instead, the cavity mode traveling downstream interacts with the beam a second time, feeding power back into the electrons; essentially this signifies a particle-accelerator interaction.
In simulations this is visible only by inspecting the electron macro-particle efficiencies on their trajectories through the geometry. Concerning the experimental situation, the relatively small power degradation caused by static ACI is not the determining observation. It is rather a limitation to the collector depression voltage during CW operation, caused by the previously described slow electrons.
It is important to note again that static ACI does not include a second RF-generating gyrotron interaction, so it is not detectable with RF measurements. The effect itself can rather be described as extended or distributed overbunching than as a parasitic oscillation.

[4] For several purposes, it is practical to exclude interactions of the operating cavity mode from this definition, and label them *dynamic ACI*; the motivations for this are given in the following.

[5] I.e. a gyrotron interaction which creates coherent RF radiation.

Dynamic ACI

A relatively new phenomenon which was found in self-consistent interaction simulations is the dynamic ACI, a term originally coined to distinguish it from the above static ACI [KAC+10b]. Dynamic ACI also describes a second interaction of the cavity mode with the electron beam at a downstream position. The determining difference is that the dynamic ACI is another RF-generating gyrotron interaction, extracting more energy from the beam. This leads to a second instance of the cavity mode, which is typically operating at a lower frequency due to the lower Ω_r and higher r_w in its interaction area.
Since self-consistent simulations are based on orthogonal modal expansion of the total RF fields, this creates massive numerical problems in simulations [AIS+14]:

- A strong beat pattern on the affected mode is caused, which introduces significant periodic or quasi-periodic fluctuations in the locally simulated power envelope (compare also [SCG+11]). As the typical ACI is multiple GHz below the cavity mode in frequency, the results is sub-ns beating, i.e. dynamic fluctuations of the RF envelope. This violates the basic simulation assumption that the baseband fields do not change during the electron transit time (see Section 3.2.2)

- The typical absorbing boundary conditions at the axial endings of the geometry are implemented for the case of single-frequency operation of each mode. The ACI oscillation is far away from the cavity mode's frequency, leading to unphysically increased reflection from the narrow-band boundary condition, changing the interaction conditions.

The inherent numerical problems involved led to a discussion on physical reality of these interactions, which sparked a systematic investigation of the influence of the various numerical schemes employed in the major EU gyrotron codes SELFT, EVRIDIKI, TWANG and COAXIAL [ARD+12, AIS+14]. The outcome of this study has hardened the

concept of dynamic ACI, which is also supported by the detection of matching frequencies in experiments [SCG+11]. There, however, remain large inconsistencies and controversy on the topic, for example that a dynamic ACI tends to be most prominent in single mode simulations; switching to multi-mode or quasi-PIC simulations removes or diminishes the appearance of dynamic ACIs in many cases. In dedicated studies, a systematic improvement of the boundary conditions in EVRIDIKI was formulated [Wu14, AWJ+14], which will allow a more qualified investigation of this aspect at KIT in the future.
Dynamic ACI was investigated recently in [Cho14], where the SELFT code variant SELFT-BZ was developed (compare Section 3.2.2). Among other gyrotrons, the W7-X gyrotrons are also discussed; as the focus was put on demonstrating the modified code functionality, only one operating point with experimental relevance was simulated in [Cho14]. Furthermore it should be noted that the category of "general ACI" as presented in this work is not discussed there, so only dynamic and static ACIs are treated separately.

6.1.4 Diagnosis of BCIs and ACIs

In a well equipped gyrotron test bed, the primary RF parasite diagnosis, i.e. positive information about its existence, comes from RF spectrum measurements.[6] If this method of detection is not employed, the symptoms listed in the previous Section yield the indication of a problem.
The basic challenge in addressing the situation is then the decision about whether the detected unexpected signal belongs to a BCI or an ACI: As soon as this is known, the problem can be narrowed down in location, leading eventually to a causal explanation for the emergence of the parasite and thus enabling a systematic remedy. In the light of the above classification, the key property for this decision is the probable dependence of the ACI on the cavity mode type, while the BCI must be independent of the cavity mode.

[6]The static ACI is of course an exception, as it does not generate an additional signal which could be measured.

If the cooling circuits for the beam tunnel, cavity and launcher are independent and CW operation is possible, the power balance from these components could give an indication on the source of the problem. This of course requires the parasite to exceed certain power levels, as it must be significant in comparison to the thermal background load from the nominal mode's wall losses and/or stray radiation. In industrial CW gyrotrons, the internal cooling circuits are normally inter-connected in series or parallel to reduce the number of external interfaces, so the diagnostic capabilities of this method are very limited.

In the following, parasitic oscillations as measured in multiple W7-X gyrotrons with the PSA system described in Chapter 4 are investigated. A strong focus is put on the RF spectrum, which can be measured in SP mode over a wide range of parameters with considerably less risk, effort and complications than in LP mode. A number of indicators, ranging from primary measurement results and parameter sweeps over analytical measurement data processing (i.e. collective dispersion analysis in the Brillouin diagram) to self-consistent interaction simulations, is introduced and applied.

6.2 Measurements

Without sensitive measurement equipment, the motivation for a search for parasitic oscillations often arises from initially inexplicable failure of the gyrotron to reach expected powers or efficiencies. Since standalone simulations of parasitic oscillations are normally afflicted with many uncertainties and the application of unvalidated methods, measurements are a crucial source of authoritative information on parasite occurrence and behavior.

Already the benchmarking of the cavity mode behavior in Section 3.2.2 has established the necessity for wide and systematic parameter sweeps. Such an approach also brings along the opportunity for the acquisition of secondary information, such as the position of mode transition points or starting currents. The retrieval of these can, however, be very time consuming and requires efficient measurement methods.

In the following, the aforementioned path of analysis is made for parasitic oscillations in the W7-X gyrotrons, moving from the level of detail to more abstract and summarizing views. After an introductory example on the spectral appearance of the parasites and their behavior during a pulse, two systematic sweeps over different external parameters are presented and discussed.[7] In the end, a summary of all relevant parasites measured in the W7-X SN7 gyrotron is presented in the form a cumulative scatter analysis of more than a hundred operating points.[8]

6.2.1 Intra-pulse behavior and spectral appearance

An important measurable quality of parasitic oscillations is their spectral appearance and behavior during a single pulse. Already in the past, it was indicated that the parasites found in the W7-X gyrotrons after the beam tunnel improvements[9] are well-defined in frequency, but do jump between different constant frequencies in the course of a single pulse lasting a few milliseconds [SCG+11].
In Figure 6.1 the spectral behavior of several observed parasitic frequencies in the context of nominal cavity mode operation is plotted. Only excerpts containing non-noise information were selected for overview reasons, so the spectrogram in the figure consists of four disjunct sub-spectrograms. The mentioned conjecture about a jumping of otherwise frequency-stable parasitic oscillations is easily validated with only one PSA measurement: Discrete parasitic frequencies are visible, which jump quasi-instantaneously.[10]
As before in Figure 5.1, the start-up tuning of the cavity mode over the external V_{cath}-I_{b}-ramp is visible, ending immediately at reaching the

[7] These sweeps were made with the primary goal of clean parasite investigation, so they do not cover the peak power operating points of the gyrotrons.

[8] The systematic investigations in this Chapter were mainly made with SN7 for practical reasons, e.g. teststand time, but they agree well with the results from other tubes of similar design such as SN6, SN2i or SN5R2.

[9] Compare [GDF+10].

[10] Taking into account the typical time resolutions used in the measurement, this means they jump faster than 1 µs.

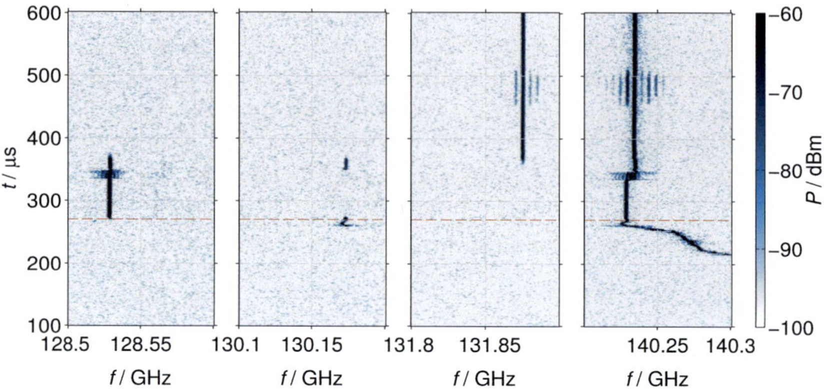

Figure 6.1: Excerpts of spectrograms showing the cavity mode and several parasites, gathered while testing W7-X gyrotron SN7 at $V_{\mathrm{cath}} = 85.3\,\mathrm{kV}$, $I_{\mathrm{b}} = 42\,\mathrm{A}$, $\alpha = 1.4$, $P = 750\,\mathrm{kW}$ with $\delta f = 600\,\mathrm{kHz}$ and $x_{\mathrm{OV}} = 0.3$

flat-top, which is indicated by the dashed line. Two instances of low-frequency modulation and a small frequency jump of the cavity mode are discernible, to which interestingly a corresponding signature in the parasites can always be found: The LFM in $340 \lesssim t_{\mathrm{p}}/\mathrm{\mu s} \lesssim 350$ is visible in the cavity mode as well as in the $128.53\,\mathrm{GHz}$ parasite. At the end of the short LFM burst, a second parallel parasite at $130.17\,\mathrm{GHz}$ becomes active, which also already had a short appearance at the end of the V_{cath}-I_{b} ramp. Soon, this parasite and the one at $128.53\,\mathrm{GHz}$ vanish and are replaced by another parasite at $131.875\,\mathrm{GHz}$. This parasite again exhibits concurrent LFM with the cavity mode in the time frame $450 \lesssim t_{\mathrm{p}}/\mathrm{\mu s} \lesssim 510$.

One important property observed in all W7-X gyrotron parasite investigations in the frame of this work is visible in the figure: The relevant parasite activity begins only approximately at the start of the HVPS flat-top. This behavior was also found in wide parameter sweeps in the $V_{\mathrm{cath}}/I_{\mathrm{b}}$ plane; it thus cannot be explained simply as an indication of starting current of the parasitic oscillation, or high minimum W_{k}.

To further illustrate typical parasite behavior, another set of spectrogram excerpts from an operating point with lower current, the parameters at $V_{\mathrm{cath}} = 88.0\,\mathrm{kV}$ and $I_{\mathrm{b}} = 38\,\mathrm{A}$ is provided in Figure 6.2. In this example, the cavity mode is not plotted since no LFM or frequency jumps were observable, only nominal behavior as already documented in Figure 5.1. The start-up ramp ended at $t_{\mathrm{p}} = 100\,\mu\mathrm{s}$, coinciding with the plotted frequency axis.
At the time flat-top is reached, a very short stub at $131.09\,\mathrm{GHz}$ is visible in the middle spectrogram, which is quickly replaced by a stable medium-power parasite at $130.81\,\mathrm{GHz}$ for $100\,\mu\mathrm{s}$, and afterwards a comparably strong one at $250 < t_{\mathrm{p}}/\mu\mathrm{s} < 340$ at $130.64\,\mathrm{GHz}$. At $t_{\mathrm{p}} = 340\,\mu\mathrm{s}$, the parasitic operation shifts to higher frequencies, with parallel activity of a dominant parasite slightly above $132\,\mathrm{GHz}$ and a weaker one at $131.03\,\mathrm{GHz}$. Both parasites last until the end of the pulse at $t_{\mathrm{p}} = 1080\,\mu\mathrm{s}$.

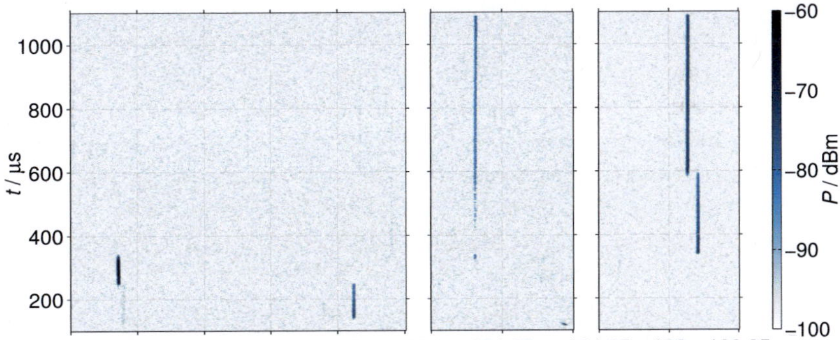

Figure 6.2: Excerpts of a spectrogram with typical parasites observed during testing the W7-X gyrotron SN7 with $\delta f = 1\,\mathrm{MHz}$ and $x_{\mathrm{OV}} = 0.2$; the external V_{cath}-I_{b} ramp flat-top lies at the lower edge of the time axis at $t_{\mathrm{p}} = 100\,\mu\mathrm{s}$, and the active cavity mode is not included in the plot.

In the two presented Figures 6.1 and 6.2 a number of properties shared by all parasitic oscillations investigated in this Chapter are exemplified:

1. Parasites only appear at (or near) the V_{cath}-I_{b} flat-top.
2. Spectral "events" (e.g. LFM bursts, frequency jumps) often appear simultaneously in the parasite and the cavity mode.
3. Some parasites appear in parallel, normally one being dominant.
4. Discrete jumping with sub-microsecond speed is often observed.
5. Between jumps, parasites have a clearly defined and constant frequency, and typically no modulation.

This emphasizes the importance of the measurement system features listed in Section 1.2.5, as both time-dependence as well as good frequency resolution are required to investigate this behavior properly. Also, without unambiguous RF frequency identification in every single shot, these frequencies would be almost impossible to measure.
Concerning the individual jumps of the parasite frequencies, due to the wide jump frequency ranges it is obvious that these cannot be performed by a gyrotron interaction maintaining fixed axial position *and* fixed mode. Thus, either mode switching occurs at a fixed z position of excitation, or the interaction jumps along the z axis, changing the mode or not. This, however, cannot be separated on the basis of isolated measurements; an attempt to do so by fusing measurement data in the Brillouin diagram is made in Section 6.3.
Still, some of the observations listed above can be used to address the basic decision between BCIs and ACIs. Point 1 yields a strong indication: The parasites behave uniformly as described at many different operating points, which were part of wide parameter sweeps in multiple dimensions. If the parasite start simply depended on a minimum I_{b} or a kinetic energy, then during the according parameter sweeps, pulses should be found which feature parasite starts during the ramp-up. Since this did not occur, it is more plausible that it is not the V_{cath}–I_{b} flat-top itself, but actually the cavity mode which is

required by the parasite. This indicates dynamic or general ACI, depending on the actual parasitic mode, following the classification in Section 6.1.

The points 2 to 4 are important information in themselves, but they do not help as much as one would believe for this question. In particular point 2 is ambiguous, as it only expresses coincidence of events. As long as the prime physical cause of the LFM bursts and the frequency jumping is not known, no fixed hierarchy between the documented effects can be formulated.

The clean spectral appearance of the parasitic frequencies noted in point 5, however, is worth a discussion of this matter. In Figure 6.3 the linewidth of a typical parasite at 130.624 GHz is compared with the nominal cavity mode at the same point of time in the pulse, using a fine frequency resolution of 10 kHz. Interestingly, the parasite has a line width no wider than the cavity mode. Very weak sidebands at ± 5 MHz are discernible; however, since they are more than 40 dB below the parasite's receiver level, their significance is small.[11]

Based on resonance condition considerations as in Section 2.2.3, the linewidth can be expected to differ for forward and backward waves: A shift in a beam line's slope or intercept will have a stronger effect on backward waves than on forward waves.[12] The narrow linewidth thus indicates rather a forward wave; this moves the focus away from the beam tunnel, which is also not likely to support such well-defined oscillations because of the inhomogeneous structures there. However, it must be noted that some resonant modes in the beam tunnel with quality factors in the range of 150 have been found in theoretical studies [Che14]. Consequently, narrow-line oscillations in the beam tunnel cannot be ruled out entirely at the time being.

During all of the studies of W7-X gyrotrons presented in this Chapter, the parasite's linewidth was always similar to the cavity mode's.

---

[11]In the Figure, the parasite even appears to have a slightly smaller linewidth than the cavity mode. This is not the case; the effect results from the existence of weak LFM at a few kHz, which is more pronounced for the cavity mode.

[12]Compare Fig. 2.2, and also Fig. 4 in [GDF+10], where the measured spectra of alleged BCI backward waves are shown.

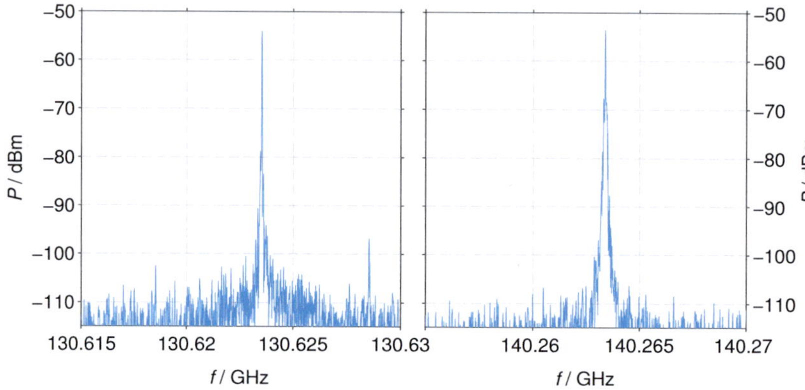

Figure 6.3: High-resolution linewidth comparison between parasite (left) and nominal cavity mode (right) at identical pulse time, during the same pulse

6.2.2 Sweep over cathode voltage

The typical sweep for parameter space exploration is a V_{cath} sweep. Despite its efficiency in covering a wide power and frequency tuning range, this type of sweep has limited informativeness for the investigation of parasitic oscillations, because most beam parameters are changed during the sweep: Increasing V_{cath} changes W_{k}, hence γ and thus also Ω_{r}. At least for diode-type guns, also α is strongly influenced by V_{cath}. Further, if no further precautions are taken, I_{b} is changed due to the Schottky effect (compare Section 2.4).

Figure 6.4 displays the relevant parameters and measured data from a V_{cath} sweep in the W7-X series gyrotron SN7, which was made at a cavity magnetic flux density of $B_{\mathrm{cav}} = 5.603\,\mathrm{T}$ with $I_{\mathrm{m}} = 84.0\,\mathrm{A}$ and $I_{\mathrm{tr}} = 4.0\,\mathrm{A}$. The measurement was conducted specially for the investigation of parasitic oscillations, so relatively high α values are reached, while the operating point is not optimal concerning output power or efficiency. Over the heater current, the Schottky effect was compensated as far as possible, for $I_{\mathrm{b}} \approx \mathrm{const} \approx 44\,\mathrm{A}$.

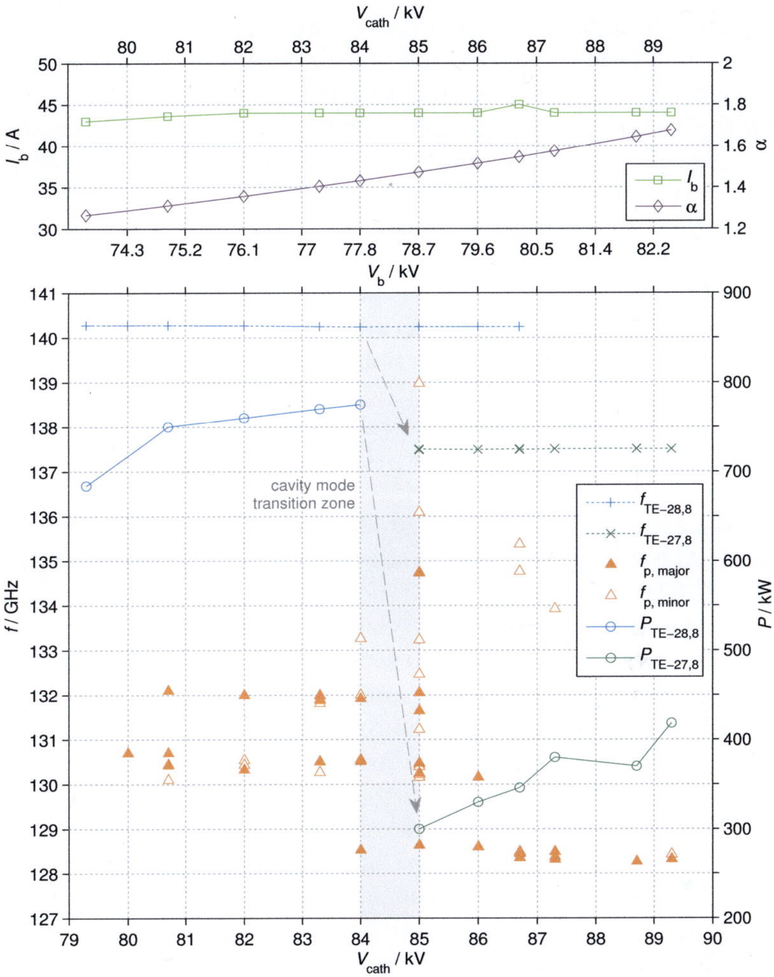

Figure 6.4: V_{cath} sweep with $I_{\text{b}} \approx$ const for W7-X tube SN7
Top: Beam current and pitch factor
Bottom: Cavity mode frequencies, measured powers and PO frequencies (filled markers indicate major parasites).

The top part of the plot indicates the extent to which I_b was kept constant, and the variation of α as a function of the applied cathode voltage. On the same axis, the bottom plot contains the measured frequencies of the cavity modes and their powers.[13]
For the investigation of parasitic oscillations, their frequency dependence on the voltage is plotted. During many pulses, multiple parasitic frequencies were measured, as was already illustrated in Section 6.2.1. To retain as much information from the spectrograms as possible, the parasites were loosely categorized into *major* and *minor*, depending on their relative time of activity during the pulse and their receiver power. In the diagram, as well as in the plots of parasite frequencies as a function of a parameter in Section 6.4, the major/minor property is conveyed over marker filling,

Observations

For $V_{\mathrm{cath}} < 80\,\mathrm{kV}$, in this configuration no parasitic activity was measured. Above this value, two bands of parasites in the frequency range 130–131 GHz and around 132 GHz can be discerned. The bands seem to tune towards lower frequency with sinking Ω_r, but the existence of multiple frequencies during the same pulse complicates analysis.
Closer to the transition to the azimuthal neighbor mode $\mathrm{TE}^-_{27,8}$, parasite activity increases with two further frequencies at 133.2 GHz and 128.5 GHz. Approaching the cavity mode transition, the gyrotron behaves quite unstably and not very reproducibly, and is very sensitive to minor fluctuations in the external parameters. A large number of different frequencies is observed, and the ones corresponding to both cavity modes in parallel often are also seen.
Above the transition voltage, output power is significantly reduced. The frequency corresponding to the nominal cavity mode can, however, still be measured over the whole duration of the pulse, albeit with ~20 dB reduced receiver power. This indicates that while the $\mathrm{TE}^-_{27,8}$

[13] Note that measured powers of the neighboring mode $\mathrm{TE}_{27,8}$ are only of limited value because the launcher is not matched to this mode, and the output window has a higher reflectivity at its frequency.

mode dominates the cavity, the nominal $TE^-_{28,8}$ mode is still present in a "lingering" fashion. Only 2 kV above the transition voltage for dominance of the $TE^-_{27,8}$ mode, a signal around 140.25 GHz can no longer be detected.
The parasite spectrum changes significantly with the cavity mode transition, since the two previously identified bands end, and a new major band around 128.5 GHz emerges.

Discussion

Most parasite frequencies appear alternating instead of in parallel, thus a discrete switching process can be assumed. Importantly, frequency bands can be distinguished, indicating that there might be groups sharing common properties, such as the interacting position and/or mode.
The behavior in the cavity mode transition zone and few kW above it clearly show that the "mode switch" terminus can easily raise a wrong impression, as the two cavity modes seem to coexist in a certain parameter range.[14] This "fuzziness" of the cavity mode transition also makes it harder to investigate the dependence of the parasites on the cavity mode.
As pointed out earlier in Section 6.1.3, in principle, the cavity mode transition could be used as a strong indicator for the interaction position of the parasitic oscillation with respect to the cavity.[15] Indeed, the sweep shows a strong dependence of the parasites on the dominant cavity mode: The observed bands are discrete except for two points, 128.517 GHz at $V_{\mathrm{cath}} = 84\,\mathrm{kV}$ with $TE^-_{28,8}$ and 130.151 GHz at $V_{\mathrm{cath}} = 86\,\mathrm{kV}$ with $TE^-_{27,8}$. The latter, however, could easily be explained by the "lingering" existence of the nominal mode up to 86.7 kV in the case of an ACI. A more definite cavity mode transition than the one found would be helpful; but the observed behavior is still pronounced enough to separate two cases:

[14]Compare also the considerations in Section 2.4.1.
[15]This would be even more the case, had the reproduction attempts in simulation (compare Chapter 3) succeeded.

- Either the parasites are interacting inside or downstream of the cavity, and depend on the bunching caused by the cavity mode as theorized in the ACI Section 6.1.3. In this case, a change in the operating cavity mode automatically causes a different parasite to start, as another pre-bunching is then encountered.[16]
- Or, the parasites are inside the beam tunnel and switch directly due to their stability range in the beam parameters. In this case, the coincident cavity mode transition would either be a real coincidence, or even be caused by the beam tunnel parasite switch.

A possible way to decide the cases would be simulations which predict where the cavity mode transition should happen if no beam tunnel oscillations were present: If the observed transition voltage was reproduced it would be an indication of ACI, while a significant difference in transition voltage would point to the beam tunnel.
Unfortunately, the simulations already fail to reproduce the mode transition with the necessary accuracy in the benchmark of Chapter 3, weakening this path of decision.

6.2.3 Sweep of magnetic flux density

As mentioned before, a principle problem of V_{cath}-variation based investigations is the resulting change of other parameters. A less "multidimensional" sweep parameter is the cyclotron frequency Ω_0, varied over the cavity magnetic flux density B_{cav}. If done simply by tuning the main coil current I_{m}, the compression ratio and therefore α is also strongly affected. For diode-type guns such as those employed in the W7-X gyrotron series, this can be countered by carefully co-adjusting the gun field over I_{tr}; however, a minor beam radius change is unavoidable.
The benefit lies in a completely constant beam energy over the whole sweep with almost uniform α, as observable from the top graph in Figure 6.4. In the other graphs (center and bottom), the measured data

[16] The unclean band transition around the cavity mode transition would then be explained by the parallel operation of $\mathrm{TE}^-_{28,8}$ and $\mathrm{TE}^-_{27,8}$, enabling several parasites.

of this experiment are plotted with common axes descriptions in the quantities I_{m}, B_{cav} and Ω_{r}, which are allowed for by $\gamma = \mathrm{const.}$ The change of r_{b} is in the order of $0.1\,\mathrm{mm}$, while α is constant within the accuracy of the simulation (~ 0.05).
In the center diagram, the development of the cavity mode and the measured power over the sweep variables is illustrated. For these data, a constant beam current of 42 A was set. Information on the parasite behavior in the context of starting currents is located in the bottom graph. These measurements were conducted with the 4-channel PSA system, so direct monitoring of the cavity mode in parallel to the parasite activity was possible through the two independent channel pairs (compare Chapter 4).
The frequency markers for the parasites in the bottom graph are coded according to the cavity mode they accompany. Over a secondary vertical axis in the bottom graph, the starting currents of the respective frequencies are shown, coded with matching marker symbols.[17] For a given measured frequency (solid markers), the corresponding starting current is marked with an empty marker of the same shape.[18] Starting current markers of adjacent frequency points are connected with lines to clarify change tendencies. In the transition area of the cavity modes $\mathrm{TE}_{28,8}^{-}$ and very probably $\mathrm{TE}_{27,8}^{-}$ around $I_{\mathrm{m}} = 83.1\text{–}83.4\,\mathrm{A}$, the minimum beam current $I_{\mathrm{b,trans}}$ for dominance of $\mathrm{TE}_{28,8}^{-}$ mode operation is also plotted as a thick dashed line.
The investigation was primarily a Ω_0 sweep, and did not cover the entire beam current parameter space. Especially above $46\,\mathrm{A}$, the operation tended to become rather unstable and of limited reproducibility concerning the parasites. The lowest beam current value at each I_{m}, however, reliably indicates the minimum parasite starting current.

[17] It is important to note that in the bottom graph of Figure 6.5, the marker types distinguish frequency and respective $I_{\mathrm{b,start}}$, and not the major/minor property of parasites as in many other figures of this Chapter.

[18] Consider as an example $I_{\mathrm{m}} = 84.5\,\mathrm{A}$: The lowest current for which a parasite was found is $I_{\mathrm{b}} = 45\,\mathrm{A}$ (indicated by "▷", right ordinate axis), which appeared in conjunction with the nominal cavity mode and at the frequency $128.45\,\mathrm{GHz}$ ("►", left ordinate axis). Already for $I_{\mathrm{b}} \geq 46\,\mathrm{A}$ (symbol "▽"), the situation changed with the takeover of parasites at $131.065\,\mathrm{GHz}$ and $132.745\,\mathrm{GHz}$ (symbols "▼").

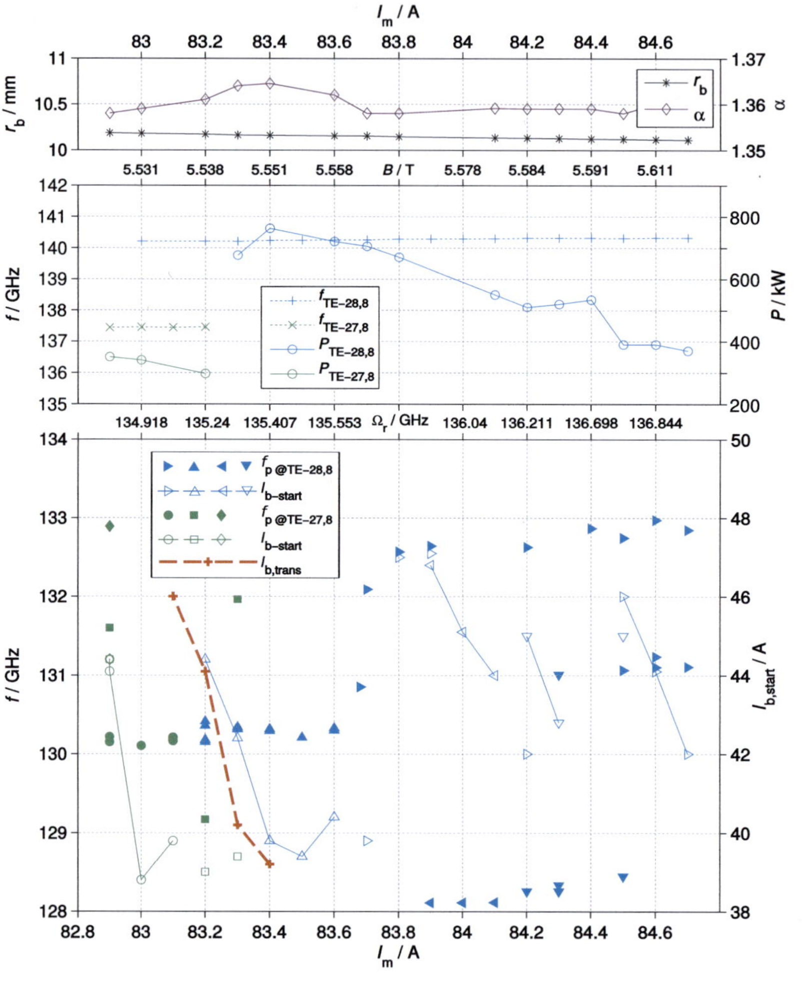

Figure 6.5: B_{cav} sweep with $V_{cath} = 82\,kV$ for the W7-X gyrotron SN7
Top: Beam radius and pitch factor
Center: Cavity modes and measured power for $I_b = 42\,A$
Bottom: Parasitic frequencies depending on the dominant cavity mode and respective starting currents

Observations

The variation of r_b and α are small throughout the used parameter range, so the experiment can be considered a clean sweep of Ω_0. For lower B_cav, the measured output power increases as expected, as the gyrotron is tuned towards higher $(\omega_\mathrm{RF} - \Omega_\mathrm{r})$ and therefore towards the "hard excitation" range. As in the previously presented V_cath sweep, no clean transition between the two main cavity modes takes place. In the range of $\mathrm{TE}^-_{27,8}$-mode dominance, a frequency corresponding to the diminished nominal cavity mode can still be measured. However, it is clear from the line levels at the receiver, the measured power and the stray radiation indicators (not plotted), that the $\mathrm{TE}^-_{28,8}$ mode does not contribute much power at these points.

Again parasitic frequency bands are visible, which alternate over the swept parameter. Unfortunately, all intervals of band-connected parasite appearance are quite short. This makes the investigation and comparison of in-band tuning directions and slopes almost impossible. The parasite starting currents strongly depend on Ω_0: While close to the cavity mode transition area at $I_\mathrm{m} = 83.4\,\mathrm{A}$, the parasites start almost together with the $\mathrm{TE}^-_{28,8}$ mode below 40 A, close by at higher main coil currents, there is parasite-free operation up to more than 46 A. Following the I_m tuning from right to left, the lowest starting currents show a pattern which resembles a sawtooth, which means a steady increase of $I_\mathrm{b,start}$ that is followed by a drop and then another gradual increase again. The intra-group starting currents follow this pattern, and tend to increase with decreasing B_cav. Comparing the difference quotients $\Delta I_\mathrm{b,start}/\Delta I_\mathrm{m}$ of the bands, a common value in the range -15 to -20 can be found. Interestingly, the respective quotient for the cavity mode is quite similar, as can be seen when comparing the inclination of the parasite $I_\mathrm{b,start}$-lines with the dashed line for the cavity mode transition current $I_\mathrm{b,trans}$.

The diagram does not directly convey over which current range a parasite mode was active. During operation with the $\mathrm{TE}^-_{28,8}$ mode, multiple parasites could be existing in parallel or excluding each other, in a manner as illustrated in Figure 6.2. However, an important difference was observed in the I_{m} range where the active cavity mode was depending on I_{b}. Crossing $I_{\mathrm{b,trans}}$ not only caused a change in the cavity mode, but also altered the parasites. Consider for example the point $I_{\mathrm{m}} = 83.3\,\mathrm{A}$, where at $I_{\mathrm{b}} = 39.4\,\mathrm{A}$ the parasite at 131.96 GHz starts in the presence of the $\mathrm{TE}^{-}_{27,8}$ cavity mode. At $I_{\mathrm{b}} = 40.3\,\mathrm{A} = I_{\mathrm{b,trans}}$, the nominal cavity mode $\mathrm{TE}^{-}_{28,8}$ takes over; no parasites were measured anymore up to $I_{\mathrm{b}} = 42.5\,\mathrm{A}$, where a 130.3 GHz parasite starts. A similar situation is encountered at 0.1 A lower main coil current at $I_{\mathrm{m}} = 83.2\,\mathrm{A}$, where for $I_{\mathrm{b}} = 39$–$44\,\mathrm{A}$ the $\mathrm{TE}^{-}_{27,8}$ mode is accompanied by a 129.15 GHz-parasite. With the mode transition at $I_{\mathrm{b}} = I_{\mathrm{b,trans}} = 44.1\,\mathrm{A}$, the parasites 129.15 GHz were not detected anymore; however, frequencies around 130 GHz immediately took over.

Discussion

Again, significant frequency band formation was found, which to some extent could be connected with the mode starting currents. This indicates again that there might be alternating modes in the same z location, or an axial jumping of the parasites. As the individual parasite frequency bands do not cover very wide uninterrupted parameter ranges, tuning tendencies (e.g. $\Delta f/\Delta\Omega_0$) inside the bands unfortunately cannot be properly determined.

The question whether the observed behavior is a result of BCI or ACI activity can again be addressed by investigating the cavity mode transition. Here the parasite starting currents in combination with the cavity mode transition current have a crucial significance; as mentioned above, both for the main coil currents $I_{\mathrm{m}} = 83.3\,\mathrm{A}$ and 83.2 A, a parasite was found which disappeared when increasing I_{b} above $I_{\mathrm{b,trans}}$ without being replaced immediately. This is a strong indicator for a dependence of the parasites on the existence of the operating cavity mode, probably due to its influence on the electron beam. This is a clear property of ACI, and excludes BCI: If the oscillations were upstream of the cavity, it would be hard to explain that for example at $I_{\mathrm{m}} = 83.3\,\mathrm{A}$ an increase in I_{b} leads to a parasite-free operation interval.

6.2.4 Scatter analysis

To gain an overview on the general distribution of parasitic frequencies, it is useful to compile a combination of all relevant measurements for a given gyrotron. A good data source are short-pulse measurements, as here very wide parameter ranges can be scanned without a significant risk of damaging the gyrotron. Also, no thermal or space charge related transient effects can be expected here; they would complicate the comparison of measured frequency data considerably.
In Figure 6.6, the detected parasites the range 125–136 GHz for the W7-X gyrotron SN7 during dominant operation of the cavity mode $TE^{-}_{28,8}$ are plotted, using data from more than 100 operating points. This macro-analysis covers the following parameter range:

- $76 \leq V_{\mathrm{cath}}/\mathrm{kV} \leq 88$
- $35.1 \leq I_{\mathrm{b}}/\mathrm{A} \leq 50$, typical parasite onset: $I_{\mathrm{b}} \gtrsim 38\,\mathrm{A}$
- $83.3 \leq I_{\mathrm{m}}/\mathrm{A} \leq 84.7$
- $1.28 \leq \alpha \leq 1.52$, below: no parasites

The visualization is nontrivial because of the many dimensions involved, which are partly interdependent. For Figure 6.6, the data are dispersed for presentation by the composite abscissa variable $W_{\mathrm{k}} \cdot I_{\mathrm{b}} \cdot \alpha$, which functions as a qualitative beam power measure.[19]
No distinct frequency dependence on this parameter is visible, however, the accompanying histogram clarifies a frequency distribution into at least three disjunct parasitic bands. The most prominent bands cover the range 130–131.1 GHz and 131.8–132.7 GHz, and contain the frequencies most often observed during experiment.[20]Further points at lower and occasionally at higher frequencies were also identified, but are considered less relevant.

[19] The data were also investigated as a function of single quantities; however no useful correlation emerged.

[20] These are also the frequency ranges in which earlier studies [SCG+11] on W7-X gyrotrons found the principal parasitic operation.

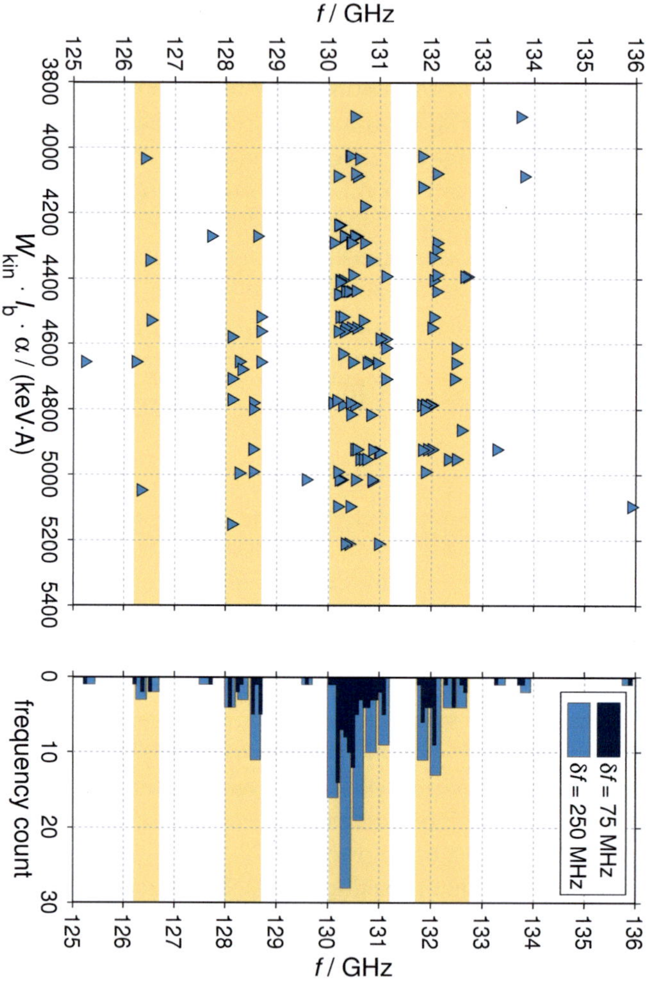

Figure 6.6: Scatterplot and histogram with two different frequency bin sizes δf of agglomerated parasitic frequencies measured in the W7-X gyrotron SN7; approximate frequency bands are denoted with yellow shading.

6.3 Brillouin-Analysis

The availability of a considerable measurement data pool calls for the use of more abstract and summarizing examination techniques, hopefully allowing larger-scale regularities or principles to become apparent. In the following, the parasitic frequencies presented in the previous Section are investigated by a modified application of the linear interaction description, which was described in Section 2.2 and applied to the cavity mode in Section 3.3.1.
This technique allows the batch analysis of measured frequencies, and yields estimations of geometrical mode excitation positions together with eigenvalue estimation. It was first described in [KSF+09, Sch09] and [SFG+10], where the first experiences were made with this technique during the investigation of BCI problems.
In this Section, improvements concerning the beam parameter calculations and the analysis technique itself are presented, whose current implementation is documented in [Kam14].

6.3.1 Fundamentals

In Section 2.2, the simplified resonance condition for an interaction between a given waveguide mode with an electron beam of defined parameters was presented, which assumes the interaction to take place at intersection points of the waveguide mode hyperbola (2.16) with the beam lines (2.20). This uncoupled dispersion relation (UDR) locally connects the properties $\omega = 2\pi f$, r_{w} and $\chi_{m,n}$ which govern the waveguide mode propagation with the beam parameters $v_{\|}$, γ and the local axial magnetic flux density B_{z}.
The frequency and the external experimental settings of an experimentally measured parasitic oscillation are initially known, while the position of excitation z_0 and the mode eigenvalue $\chi_{m,n}$ are sought for. If the beam parameters along the axis $\Omega_{\mathrm{r}}(z)$ and $v_{\|}(z)$ are available, the UDR (2.21) defines together with the geometry profile $r_{\mathrm{w}}(z)$ a pair of $[k_{\|}, \chi_{m,n}]^i$ at each z position for each measured frequency f^i. This information becomes useful, when over a variation of external

parameters different but similar parasitic frequencies are measured: If the signals show common properties and tune monotonically over the external values, it can be tentatively assumed that these frequencies originate from interaction of one and the same mode at one fixed z-position.

The Brillouin diagram analysis is the test of this assumption using the aforementioned value pairs $[k_{\|}, \chi_{m,n}]^i(z)$, one for each measured frequency: If the assumption is true, there must be at least one axial position z_0 at which all the measured frequency points constitute a hyperbola which satisfies (2.16) in the Brillouin diagram. As the only free parameter of the hyperbola is the eigenvalue, finding the hyperbola is analogous to searching for a position z_0 where all value pairs $[k_{\|}, \chi_{m,n}]^i(z)$ share a common eigenvalue $\chi_{m,n}$.

A practical way of solving this problem is to calculate the variance of the eigenvalues along the axis $\sigma^2(z) = \mathrm{Var}\left(\chi^i_{m,n}(z)\right)$, and to investigate minima in these data. This corresponds to searching for the eigenvalue $\overline{\chi}$ of the hyperbola $f(k_{\|})$ yielding the minimum frequency difference mean-square error for the individual frequency points.

6.3.2 Beam parameters along the z axis

The z-profiles of the relevant beam parameters can be completely simulated with the codes mentioned in Section 3.2.1. However, with the adiabatic relationships from Section 2.1.2, they can also be determined from reference values with less effort: If the beam parameters are known at a reference position z_{ref}, using (2.10) together with the $B_{\mathrm{z}}(z)$ profile, these values can be analytically calculated in every other z position.

However, one should take into account that $\gamma(z)$ is determined by the voltage depression effect, which complicates the problem. Consider a known set of reference parameters $\beta_{\perp,\mathrm{ref}}$ and γ_{ref} at the reference point z_{ref}, simulated from the experimental external parameters V_{cath}, I_{b} with the complete magnetic profile $B_{\mathrm{z}}(z)$. From the voltage depression term (2.12) together with (2.11) and (2.3), the relationship

$$\frac{eI_\mathrm{b}}{2\pi\epsilon_0\beta_\| c_0}\ln\left(\frac{r_\mathrm{w}}{r_\mathrm{b}}\right) = eV_\mathrm{cath} - m_{\mathrm{e},0}c_0^2\,(\gamma - 1) \tag{6.1}$$

is obtained. r_b can be directly inserted from (2.9), leaving $\beta_\|$ and γ unknown. With the relation

$$\beta_\| = \sqrt{\frac{1 - 1/\gamma^2}{1 + \alpha^2}} \tag{6.2}$$

and the adiabatic connection to the reference data from (2.10)

$$\alpha\gamma\beta_\| = \gamma_\mathrm{ref}\beta_{\perp,\mathrm{ref}}\sqrt{\frac{B_\mathrm{z}}{B_\mathrm{z,ref}}} =: K \quad , \tag{6.3}$$

three equations for the three unknowns α, $\beta_\|$ and γ are now given. Substituting α and γ, an implicit energy conservation equation

$$\frac{I_\mathrm{b}}{2\pi\epsilon_0\beta_\| c_0}\ln\left(\frac{r_\mathrm{w}}{r_\mathrm{b}}\right) - V_\mathrm{cath} = \frac{m_{\mathrm{e},0}c_0^2}{e}\left(1 - \sqrt{\frac{K^2+1}{1-\beta_\|^2}}\right) \tag{6.4}$$

is established, which for consistent physical input parameters has exactly one non-complex solution $\beta_\|$. This way, an "unspent" electron beam can be approximately characterized in the entire geometry, using only fully known beam parameters at z_ref.

Inside the cavity and downstream, these calculations no longer hold, as the main interaction inside the cavity alters the beam parameters. However, downstream of the cavity, adiabatic behavior can again be expected, and if the cavity mode's RF power is known, there is possibility for improvement: The generated RF power in the cavity P_gen can be estimated from the measured values as described in Section 3.1.3, allowing for the application of the first-order beam parameter average correction established in Section 2.2.3 to the data in $z > z_\mathrm{cav}$.[21] In the following, χ variances based on the uncorrected beam parameters are denoted by $\sigma_0^2(z)$, to be distinguished from the power-corrected $\sigma_\mathrm{P}^2(z)$.

[21] Unlike in Section 2, here P_gen is used instead of P_diff, since the Brillouin analysis does not consider RF losses. Compared with the uncertainties inherent to the Brillouin analysis, the difference between the two values is rather small anyway.

6.3.3 Applicability and scope

The UDR was tested in Section 3.3.1 for the frequency estimation of the cavity mode, which failed to be more precise than one percent in frequency. A very probable reason lies in the strong effect of the powerful cavity interaction on the beam, weakening the validity of the simulated beam parameter values, as discussed there.[22]

Contrary to the gyrotron interaction in the cavity, parasitic oscillations can be expected to operate at a much lower power range, that is smaller than 1% with respect to the total electron beam power.[23] As a result, the primary (i.e. average) parameters of the electron beam are not changed significantly by a PO. Thus, the linear calculations can be expected to be more applicable to these oscillations.

It must be made clear that this method is only able to give an indication, and not final information on the excitation position and the mode of a parasitic oscillation. The main purpose might be the fundamental decision between BCI and ACI activity, which then points to subsequent methods to be applied. Nonetheless, the possibility to fuse various measurements into one batch in order to find probable interaction positions and eigenvalues from common properties between all points is a truly unique feature of this method.

The result must be interpreted as the answer to the hypothesis that all considered measurement values in the analysis come from the same mode at the same interaction position: If the analysis yields no conclusive result, then the hypothesis is not true. If, on the other hand, the method yields a definitive z position, the hypothesis is correct. Also, this event in itself justifies the method to a certain extent, as the formation of a clear result purely from chance is not very probable.

[22] The more sophisticated CDR achieved worse frequency agreement, which, however, is not the only reason why it is not considered here: For the calculation of the CDR, in addition to the eigenvalue, the indices of the mode in question must be known. This removes the possibility of simply inverting the relationship $(\chi, r_\mathrm{w}, \alpha, \gamma) \mapsto (k_\parallel, f)$ as done with the UDR, and therefore would require testing all modes separately.

[23] Compare also the discussion in Section 6.1.1.

6.3.4 Method summary

The described dispersion analysis in the Brillouin diagram can be summarized as follows:

- SP experimental parameter sweeps, with measurement of RF frequency and power
- Simulation of beam parameters at a reference point z_{ref} in the geometry
- Calculation of beam parameters along the z axis using $B(z)$ and $r_{\mathrm{w}}(z)$ with the relations listed in Section 6.3.2, solving the implicit equation (6.4) numerically at each z point
- Power correction to the beam parameter values in $z > z_{\mathrm{cav}}$
- Calculation of the Brillouin data $[k_{\parallel}, \chi_{m,n}]^i(z)$ for each frequency point of the sweep
- χ variance calculation for each z value, yielding $\sigma_0^2(z)$ and $\sigma_{\mathrm{P}}^2(z)$
- Assessment of $\sigma^2(z)$ minima in the Brillouin diagram.

If the selected frequency points indeed fulfill the assumption of common z_0 and $\overline{\chi}$, the method yields a probable combination of these values, and additional information about the nature of the parasitic oscillation through $k_{\parallel}^i$.

In a following step, it is useful to evaluate the beam-wave coupling coefficients of the TE modes with $\chi_{m,n} \approx \overline{\chi}$ at z_0, as the Brillouin diagram only considers the uncoupled resonance condition. For the choice of the χ range, the benchmark study in Figure 3.3 can be consulted: The deviation in frequency between the measured and the calculated curves is in the range of hundreds of MHz for the lower power operating points of a cavity interaction; this could resemble the case for parasitic oscillations. For oscillations close to cut-off, $\chi_2/\chi_1 \approx f_1/f_2$; hence the allowed percentage frequency deviation from the hyperbola directly defines the χ range. Allowing, for example, a deviation of $\pm 500\,\mathrm{MHz}$ for a parasite at 130 GHz yields $\pm 0.4\%$, hence for $\overline{\chi} = 60$ a range of 59.76–60.24.

6.3.5 Investigation of the W7-X gyrotron parasites

Parasite band 131.8 - 132.7 GHz

Already in the scatterplot in Figure 6.6, at least one contiguous subband around 132 GHz in the composite axis range 4300–4600 keVA can be recognized. However, in a first attempt, the Brillouin analysis is applied to all frequency points in the entire band to show its behavior. Figure 6.7 contains the $\sigma_0^2(z)$ (hyperbola χ variance) and $\sigma_\mathrm{P}^2(z)$ (hyperbola χ variance with power correction) curves, in addition to a schematic geometry.
Minima in the $\sigma^2(z)$ curves hint at the most probable locations of interaction; it should be kept in mind that this only accounts for the selected group of points: If these points happen to correspond to interactions from different z-positions or different modes, the method will not yield meaningful results.
There are two minima, one at the beam tunnel exit at $z_0 = 325\,\mathrm{mm}$, and one in the launcher at $z_\mathrm{P} = 478\,\mathrm{mm}$. Both are not very distinct, but it should be pointed out that $\sigma_0^2(z)$ shows no sign of a minimum at z_P, so this minimum was created solely by the power correction. Because the power correction is performed individually for each point, this alone is already an indication of ACI.[24]
The Brillouin diagrams corresponding to the found minima are provided in Figure 6.8 for closer analysis. In both of them the $\overline{\chi}$ waveguide hyperbola is also indicated.
For the beam tunnel minimum (see Fig. 6.8a), the interaction points are located close to cut-off, and mostly in the slow backward wave half-plane. The structure does not show much agreement with the plotted or any other conceivable hyperbola; even groups with shapes which would fit to forward wave hyperbolas are discernible.
In Figure 6.8b, the Brillouin plot at $z_\mathrm{P} = 478\,\mathrm{mm}$ is shown. It is notable how the different assumed position and the power correction shift the individual points in the $k_\|$ dimension, when compared with

[24] Otherwise, the power correction would only arbitrarily perturb $\chi^i(z)$ for $z > z_\mathrm{cav}$, without much change in the total variance .

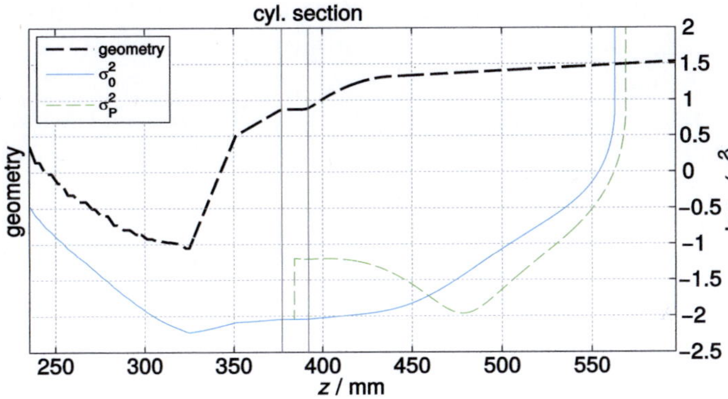

Figure 6.7: Geometry radius and variance of χ-values as a function of the z-coordinate for the parasite band 131.8–132.7 GHz (compare also Fig. 6.6)

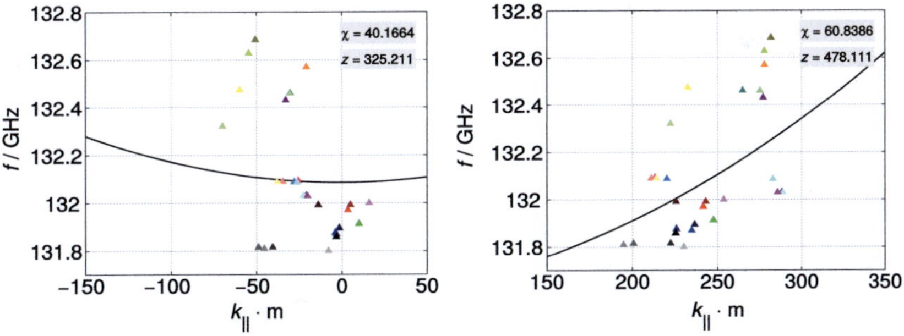

(a) σ_0^2-minimum at position $z = 325\,\mathrm{mm}$ (beam tunnel)

(b) σ_P^2-minimum at position $z = 478\,\mathrm{mm}$ (after-cavity region)

Figure 6.8: Brillouin diagrams of 131.8–132.7 GHz parasites for the two σ^2-minima in Figure 6.7

Figure 6.8a. Here a more uniform result is found, with all points indicating a moderate forward wave. Again, grouping is visible: The points below the hyperbola have a fitting inclination, while the points with $f > f_{\mathrm{hyp}}$ form two distinct sub-groups which are both more strongly inclined, i.e. have a higher group velocity.
The two major groups can be separated in order to check the behavior of the Brillouin diagram: They could stem from different z positions, but still be excited in the same mode. First, the altered variance plots for the two sub-bands ⓐ (lower frequencies) and ⓑ (higher frequencies) are investigated; these are shown in Figure 6.9. The most notable change is the expected deepening of the $\sigma_P^2(z)$ minima, combined with a z-shift. However, since the sub-bands are more homogeneous than the complete band, the $\sigma_0^2(z)$- curves also show better matching. But the z-position of their σ_0^2 -minima does not change, so the associated Brillouin diagrams for $z_{0,\mathrm{a/b}}$ will be identical to the one in Figure 6.8a. In Figure 6.10, the Brillouin diagrams for the respective ACI minima in the sub-bands ⓐ and ⓑ are plotted. Both now follow the hyperbola in a smoother way, albeit by far not as smooth as the result of the investigations on BCIs in [Sch09]. The reason might be that the individual points do not originate from the exactly the same z-position, and surely the inherent uncertainties of the linear theory as explored in Section 3.3.1 should also be kept in mind.
The found eigenvalues $\chi_{\mathrm{a}} \approx 60.87$ and $\chi_{\mathrm{b}} \approx 60.76$ now allow the investigation of the local beam-wave coupling coefficients *Ge* (2.18). As the two minima ⓐ and ⓑ are adjacent with respect to χ and z, they can be investigated together. Each frequency point f^i of the batch represents an own set of beam parameters; thus, also the individual coupling factor curves $Ge(z)$ are different for each point. This allows another layer of investigation: Modes with a high $Ge^i(z_{\mathrm{a/b}})$ for all considered points i are better candidates than modes which only couple well to a part of the batch. To include this information in all $Ge(z)$ plots, i.e. the Figures 6.11 and 6.14, the maximum and minimum relative coupling coefficients Ge^{r} as defined in (3.10) for the identified mode candidates are plotted.

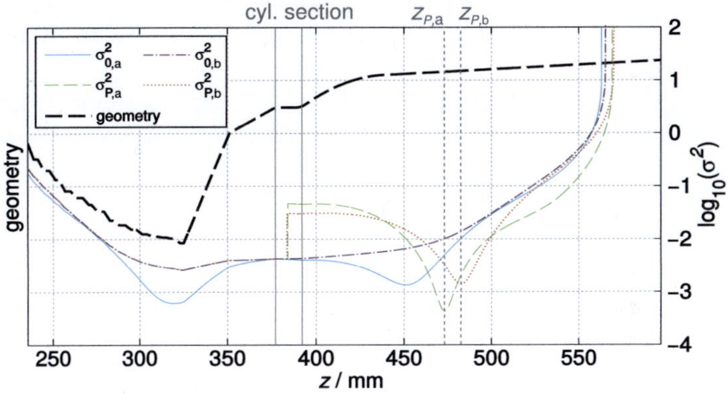

Figure 6.9: Geometry radius and variance of χ-values over the z-coordinate for the parasite sub-bands ⓐ 131.8–132.05 GHz and ⓑ 132.07–132.7 GHz, using original beam parameters σ_0^2 and the power-corrected version σ_P^2

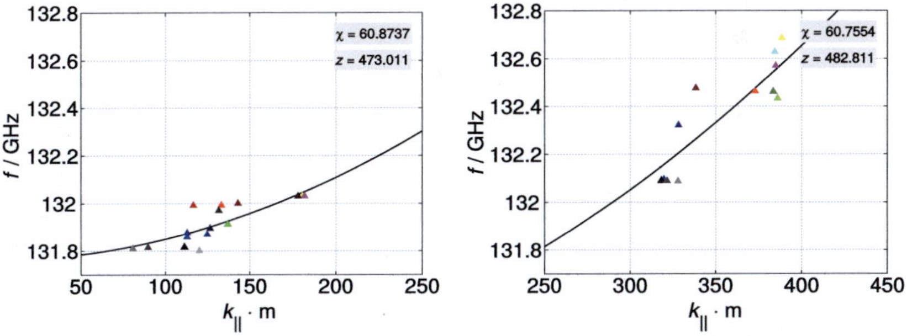

(a) σ_P^2-minimum of 131.08–132.05 GHz sub-band at $z_{\mathrm{P,a}} \approx 473.0\,\mathrm{mm}$

(b) σ_P^2-minimum of 132.07–132.7 GHz sub-band at $z_{\mathrm{P,b}} \approx 482.8\,\mathrm{mm}$

Figure 6.10: Brillouin diagrams of the split parasite sub-bands for the σ_P^2-minima in Fig. 6.9

Based on these estimates, the range $60 < \chi < 61.4$ was investigated. Of the 24 transverse-electric modes present in this χ range, five have a relative coupling $Ge^{\mathrm{r}}(z_{\mathrm{a}}) > 0.2$ or $Ge^{\mathrm{r}}(z_{\mathrm{b}}) > 0.2$, as can be seen in Figure 6.11.

Interestingly, at both positions the $\mathrm{TE}_{28,8}$ mode has a very high coupling value, which indicates dynamic ACI. With $\chi_{28,8} \lesssim \overline{\chi}$, it also satisfies the lower-edge-of-excitation-band consideration, and shows the smallest ΔGe^{r}-range at z_{a} and z_{b}. Other good candidates are the $\mathrm{TE}_{29,8}$ and $\mathrm{TE}_{26,9}$ modes, which are typical competitors (compare Table 3.2 and Section 2.4.1). They represent potential cases of general ACI; however, with $\chi_{29,8} \approx 61.33$ and $\chi_{26,9} \approx 61.13$ their eigenvalues are above the ones found.

The remaining two candidates $\mathrm{TE}_{13,14}$ and $\mathrm{TE}_{18,12}$ are rather exotic modes, which are rarely encountered in simulations, as they also have rather low coupling in the cavity area.

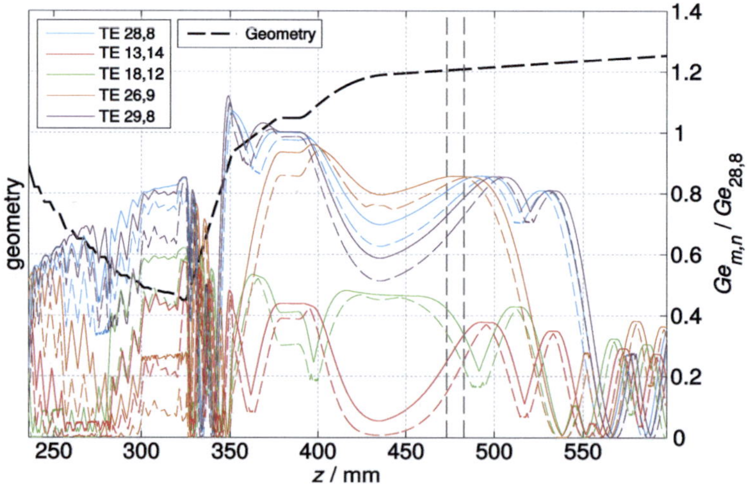

Figure 6.11: $Ge^{\mathrm{r}}(z)$ of the modes with maximum Ge^{r} at $z = 483\,\mathrm{mm}$ and $z = 473\,\mathrm{mm}$ with $60 < \chi < 61.4$ (compare Fig. 6.9; solid lines: maxima, dashed lines: minima)

Parasite band 130.5 - 131.2 GHz

As another example, a part of the main parasite band slightly above 130 GHz is investigated in the same fashion as above.
The variance plot is found in Figure 6.12, and shows behavior similar to the ones from the other band: σ_0^2 exhibits a weak minimum in $300 < z/\mathrm{mm} < 310$, and a secondary weak minimum at the launcher entrance, which is considerably deepened by the power correction. The global σ_P^2-minimum is found at $z_2 \approx 487\,\mathrm{mm}$, not far from the ones in the previous band.
A plot of the corresponding Brillouin diagrams is given in Figure 6.13, where again the beam tunnel minimum (Fig. 6.13a) shows poor matching. The complete group resides in the slow forward wave half-plane, but especially in the range $90/\mathrm{m} < k_\| < 120/\mathrm{m}$ there is a sub-group of points clearly requiring a higher group velocity at this z than the dispersion hyperbola grants. Also, a number of outlier points clearly have no relation to the rest of the batch.
The σ_P^2 minimum in the after-cavity region, as plotted in Figure 6.13b, shows a much better correspondence between the hyperbola and all the measured frequency points. Interestingly, the found wave number $k_\|$ is similar to the one found for the parasites in the frequency range 132.07–132.7 GHz (compare Fig. 6.10b), in the slow forward wave range. Note also how the drastic outliers in Figure 6.13a were smoothly integrated into the batch by the power correction.
This encourages another coupling coefficient investigation in the after-cavity region, albeit with a slightly lower eigenvalue range than before, using $59.4 < \chi < 60.6$. Of the 19 TE modes existing here, only four are significant as far as $Ge^\mathrm{r}(z_\mathrm{P})$ is concerned. The axial dependence $Ge^\mathrm{r}(z)$ is shown in Figure 6.14. Again, a grouping into two dominant modes, $\mathrm{TE}_{28,8}$ and $\mathrm{TE}_{25,9}$, and two rather minor modes, $\mathrm{TE}_{17,12}$ and $\mathrm{TE}_{10,15}$, around z_P emerges. In addition to their unusual index combinations and low $Ge^\mathrm{r}(z_\mathrm{P})$, the latter group also exhibits larger $\Delta Ge^\mathrm{r}(z_\mathrm{P})$.
Peculiarly, the two dominant modes found, $\mathrm{TE}_{28,8}$ and $\mathrm{TE}_{25,9}$, are the same as for the previous batch of points (compare Section 6.3.5 and especially Fig. 6.9).

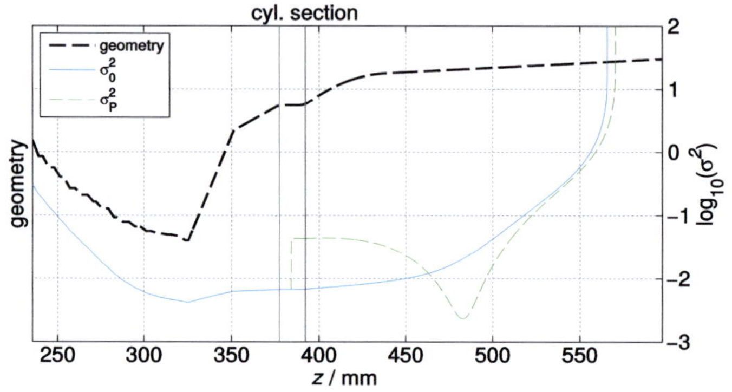

Figure 6.12: Geometry radius and variance of χ-values over the z-coordinate for parasites in the band 130.5–131.2 GHz, comparing directly calculated beam parameters (σ_0^2) and power-corrected version (σ_P^2)

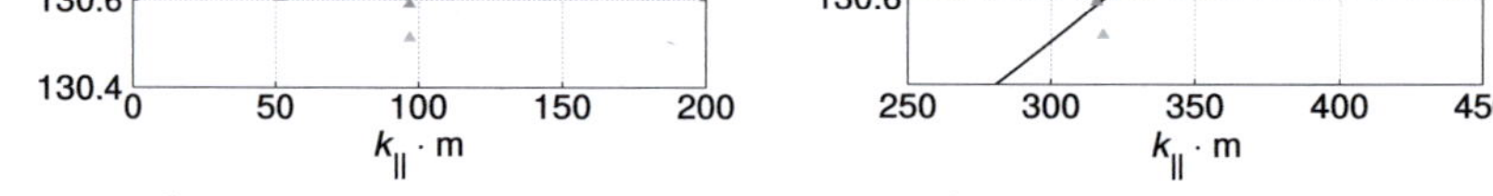

(a) σ_0^2-minimum at $z_0 \approx 307$ mm

(b) σ_P^2-minimum at $z \approx 487$ mm

Figure 6.13: Brillouin diagrams of the 130.5–131.2 GHz parasites at both found σ^2-minima (compare Fig. 6.12)

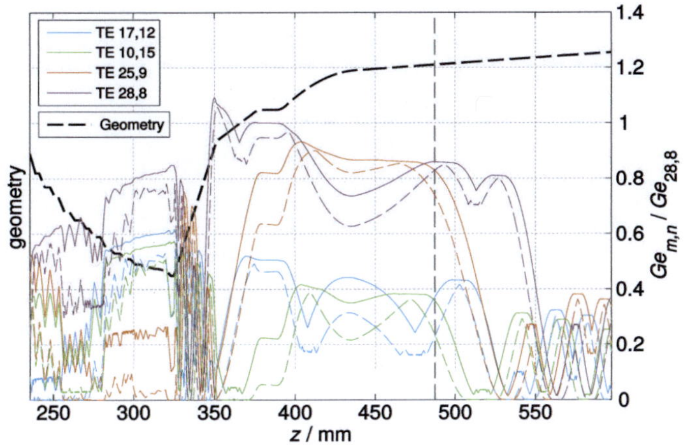

Figure 6.14: $Ge^{\mathrm{r}}(z)$ of the modes with maximum Ge^{r} at $z = 487\,\mathrm{mm}$ with $59.4 < \chi < 60.6$ (compare Fig. 6.12)

6.3.6 Result summary

Apart from the examples presented, more Brillouin investigations were made with the measured parasite frequencies. A summary of the results obtained frequencies in the upper two frequency bands marked in Figure 6.6 is compiled in Table 6.1.The Brillouin analysis in the range of the 128 GHz band yielded no conclusive result, indicating that these frequencies do not come from the same mode interacting at one position.

The most fundamental result of the analysis was already stated, which is the fact that the application of the cavity mode power extraction correction in each of the named bands lead to the formation of a global and well defined minimum of $\sigma^2(z)$ in the launcher region. This strongly indicates an ACI phenomenon, and shows that the beam parameter correction introduced to the method was necessary.

sub-band	z	$\overline{\chi}$	$\mathrm{TE}_{m,n}$	$\chi_{m,n}$
$131.8 - 132.5$	473	60.87	$13, 14$	60.64
			$18, 12$	60.93
$132.07 - 132.7$	483	60.76	$28, 8$	60.10
			$26, 9$	61.13
			$29, 8$	61.33
$130.5 - 131.2$	487	60.08	$17, 12$	59.6
			$10, 15$	59.63
			$25, 9$	59.88
			$28, 8$	60.10
$130.2 - 131.2$	476	60.17	$25, 9$	59.88
			$28, 8$	60.10

Table 6.1: Frequency bands of successfully investigated batches, obtained z-positions with average eigenvalues; modes from *Ge* investigations and corresponding eigenvalues

Concerning the mode estimation, the investigated frequency bands share a number of properties:

- All matches are found in a relatively compact area, 473–487 mm, only a few centimeters into the launcher. As a gyrotron interaction also requires a certain axial length, and the extent of the matching area is of the same order of magnitude as the cavity's, this might in fact simply be one interaction zone, which however, possibly due to the helical wall perturbations, has less defined borders than the cavity.

- The nominal cavity mode $\mathrm{TE}_{28,8}$ appears in all successful matches. Depending on the actual rotation direction, following the definitions made in Section 6.1, this either indicates dynamic ACI ($\mathrm{TE}_{28,8}^{-}$) or general ACI ($\mathrm{TE}_{28,8}^{+}$).

- Another mode which is found as a top candidate in several matches is $\mathrm{TE}_{25,9}$, which is known to be a relevant competitor inside the cavity itself.

- Without exception, the interactions in the Brillouin diagram at the matching z positions constitute forward waves relatively close to cut-off ($k_{\|} < 500/\mathrm{m}$). This corresponds to ordinary gyrotron oscillator interactions, and agrees with the linewidth considerations in Section 6.2.1.

The further matches include the possibly relevant modes $\mathrm{TE}_{26,9}$ and $\mathrm{TE}_{29,8}$, and some rather exotic candidates.
Concluding, it must be noted that while the Brillouin diagram matches are reasonable, they are unfortunately not as striking as for the case of BCIs in [Sch09, SFG+10], despite advancement of the method and the availability of a large data pool for the investigation presented in this work.[25]
Since the interpretation of the measurement sweeps and, even more so, the Brillouin analysis results indicate after-cavity interactions, the application of the simulation techniques reviewed and benchmarked in Chapter 3 is a logical next step.

6.4 Comparative simulations

6.4.1 Simulation setup

In order to investigate ACI phenomena, a part of the launcher must be included in the simulation. This, however, is problematic, as simulations become increasingly unstable with extension of the geometry. As a compromise between the prediction of the Brillouin analysis in the previous Section and the limitations of the simulations, 40 mm of launcher were added after the end of the nonlinear uptaper (compare Fig. 6.15a). In reality, the mode-converting corrugations in the launcher surface provide a complicated boundary condition for all modes, inducing mode-conversion, which cannot be modeled by the inherently axisymmetric interaction codes.[26] Simulations in this area

[25]Of course by selectively reducing the number of points, better fits could have been achieved. It is, however, problematic when outliers are removed a-posteriori to improve results, so this was not done.
[26]See also Section 3.2.2 on the limitations of the simulation tools.

therefore assume different conditions for the waveguide modes, e.g. ignoring mode conversion and scattering effects. The applicability for parasite studies is limited by this, as this could erroneously enhance or suppress interactions in this region; probable consequences include overall instability, significantly different parasite starting currents and possibly altered mode spectra.
In contrast to the basic simulation study in Chapter 3, where only cavity interactions were of interest, now the complete geometry must be analyzed for the mode list. The selection criteria were set as follows:

- $Ge^{\mathrm{r}}(z) > 0.6$; this means the local coupling coefficient normalized to the $\mathrm{TE}_{28,8}$ mode coupling in the cavity midsection, as defined in (3.10).
- $128 \leq f_{\mathrm{cut}}/\mathrm{GHz} \leq 155$ inside the cavity; for the $\mathrm{TE}_{28,8}$ mode, this translates into $119.7 - 145\,\mathrm{GHz}$ at the uptaper end, so a sufficiently wide frequency range is covered.
- $350 < z/\mathrm{mm} < 480$; this excludes most the cut-off section, which, due to its large radius change, would strongly increase the number of modes to be considered.

With these values, a list of 80 modes is obtained. Except for the modes with azimuthal indices below 18, all modes suggested by the Brillouin analysis are part of this initial selection. The exotic modes $\mathrm{TE}_{13,14}$ and $\mathrm{TE}_{18,12}$ were added manually, yielding a final 84 modes (see Table 6.2). The coupling coefficient spectrum is illustrated with $Ge^{\mathrm{r}}\left(f_{\mathrm{cut}}(z_{\mathrm{cav}})\right)$ in Figure 6.15b.
The extraordinarily large mode list places a considerable burden on any interaction simulation code, and without capable codes and adequate parallel computing resources it would be impossible to apply it. Because of this and the experiences made with the benchmark in Chapter 3, the simulation codes EVR-BZ and EVR-PIC were chosen for this task. It should be noted that this is relatively new ground, as typical multi-mode interaction simulations employ at most one or two dozen modes. This study thus also investigates the consequences of very extensive mode lists on the simulation codes and the related tools.

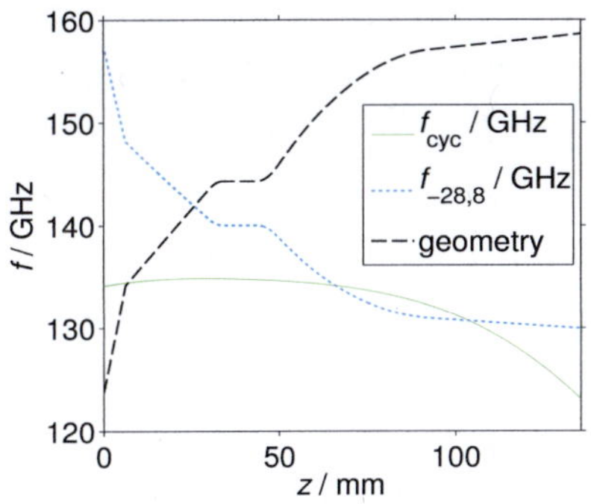

(a) Geometry, relativistic electron cyclotron frequency f_{cyc} and $\mathrm{TE}_{28,8}$ mode cut-off frequency $f_{-28,8}$

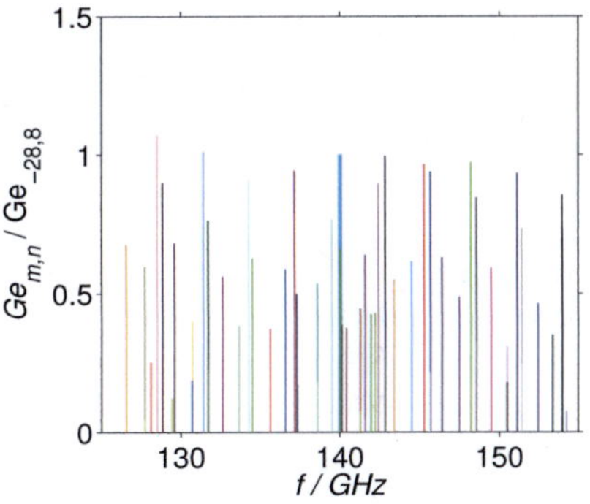

(b) Relative coupling inside the cylindrical cavity section versus cut-off-frequency

Figure 6.15: Simulation scenario overview

$-11,15$	$+19,11$	$+22,10$	$+24,9$	$+26,9$	$\mathbf{-28,8}$	$-30,9$
$+11,15$	$-20,10$	$+22,11$	$-24,10$	$-26,10$	$+28,8$	$+30,9$
$-13,14$	$+20,10$	$+22,12$	$-24,11$	$+26,10$	$-28,9$	$-31,7$
$+13,14$	$+20,11$	$-23,9$	$+24,11$	$-27,7$	$+28,9$	$-31,8$
$+14,12$	$-21,9$	$+23,9$	$-25,8$	$+27,7$	$-29,7$	$+31,8$
$-16,11$	$+21,9$	$-23,10$	$+25,8$	$-27,8$	$-29,8$	$-32,7$
$+18,10$	$-21,10$	$+23,10$	$+25,9$	$+27,8$	$+29,8$	$-32,8$
$+18,11$	$+21,10$	$-23,11$	$-25,10$	$-27,9$	$-29,9$	$+32,8$
$-18,12$	$+21,11$	$+23,11$	$+25,10$	$+27,9$	$+29,9$	$-33,7$
$+18,12$	$-22,9$	$-24,8$	$-26,8$	$+27,10$	$-30,7$	$-33,8$
$-19,10$	$+22,9$	$+24,8$	$+26,8$	$-28,7$	$-30,8$	
$+19,10$	$-22,10$	$-24,9$	$-26,9$	$+28,7$	$+30,8$	

Table 6.2: TE mode list for parasite simulations

The start-up ramps are configured as in the benchmark study to start from $V_\mathrm{b} = 25\,\mathrm{kV}$ and reach the final point after $2\,\mu\mathrm{s}$. With EVR-BZ running massively parallel using 512 threads on 32 HELIOS nodes, one simulation could be completed within less than 24 hours.
Since no severe pathological activity was found, no mode exclusions as in Chapter 3 had to be made.

Sample time behavior

Before investigating the wide parameter sweeps performed, in the following, a sample operating point is presented. Figure 6.16 contains the time evolution for the point $V_\mathrm{cath} = 73.7\,\mathrm{kV}$, $I_\mathrm{b} = 44\,\mathrm{A}$ as simulated with EVR-BZ. To increase the clarity in the graph, only modes with $P_\mathrm{sim} \geq P_\mathrm{min} = 10\,\mathrm{kW}$ are included in the legend of this graph and the related ones.
Presumably due to the very large number of considered modes, a more heterogeneous start-up process than in the benchmark studies is encountered. Almost at the start of the ramp-up, the $\mathrm{TE}^+_{26,9}$ mode starts and alternates briefly with the $\mathrm{TE}^-_{27,9}$ mode. After a short activity of the $\mathrm{TE}^+_{25,9}$ mode, the nominal mode starts and takes over quickly. During its start, some $\mathrm{TE}^+_{24,9}$ mode activity is visible, which, however, is again quickly replaced by the $\mathrm{TE}^+_{25,9}$ mode. The latter mode then also remains an active parasite in steady state, where the $\mathrm{TE}^-_{28,8}$ mode oscillates with an output power of $P_\mathrm{sim} = 800\,\mathrm{kW}$.
The field patterns as obtained in steady state at $t_\mathrm{p} = 2500\,\mathrm{ns}$ are plotted in Figure 6.17. The most striking feature is the clear ACI character of the $\mathrm{TE}^+_{25,9}$ mode. The weak residual $\mathrm{TE}^+_{24,9}$ activity during steady state, as visible in Figure 6.16 corresponds to a rather delocated field pattern, having its major part in the range of the cavity midsection. Within the background noise, a number of modes is discernible which just seem to reach cut-off at the left end of the cavity; had not a part of the cut-off section been considered in the simulation, these modes would encounter a non-reflecting boundary condition in the upstream direction, and likely behave different.

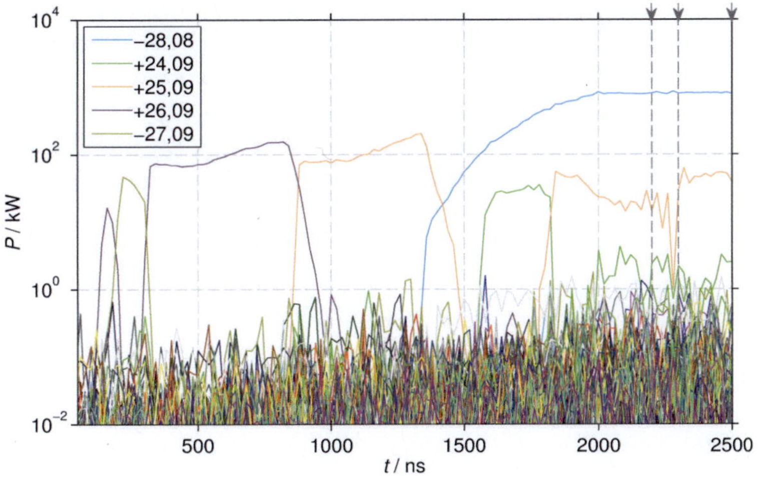

Figure 6.16: Mode evolution for $V_{\mathrm{cath}} = 73.7\,\mathrm{kV}$ with final state including parasitic $\mathrm{TE}^{+}_{25,9}$ mode activity; $P_{\min} = 10\,\mathrm{kW}$

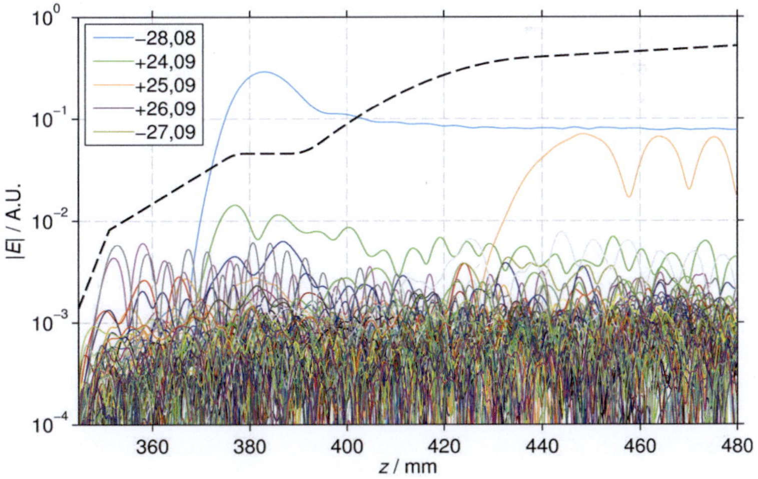

Figure 6.17: Field patterns for $t_{\mathrm{p}} = 2500\,\mathrm{ns}$ in Fig. 6.16

A number of spectra, gathered from the simulation time points[27] $t_\mathrm{p}/\mathrm{ns} = (2200, 2300, 2500)$, which are also indicated in Figure 6.16, are presented in Figure 6.18. The frequency span was reduced to the range containing significant activity, $129 < f/\mathrm{GHz} < 144$.
As it was already visible from the previous figures, the nominal cavity mode at 140.27 GHz dominates. It has no sidebands or secondary peaks, and hence no significant dynamic ACI activity.
The second important oscillation is the $\mathrm{TE}^{+}_{25,9}$ mode, which is found to be active as a general ACI around 131 GHz. At powers up to 50 kW, as visible in Figure 6.16, the simulated parasite power in this example is larger than expected from the experimental experience, but not in an unphysical way. An explanation could be the smooth launcher surface in the simulation, which can be expected to provide much better interaction conditions than a corrugated surface. This parasite was the strongest one in the considered simulation batch; typical levels in the other simulations were between 10 kW and 40 kW.
Interestingly, during steady state of the nominal cavity mode, a frequency jump can be observed in the parasite: At $t_\mathrm{p} = 2200\,\mathrm{ns}$ (top spectrum in Fig. 6.18), its main line is at 131.250 GHz, showing sidebands in the range ± 70 MHz. In the middle spectrum, which represents the situation 100 ns later (compare also the time indicators in Fig. 6.16), the lower sideband line seems to increase. At $t_\mathrm{p} = 2500\,\mathrm{ns}$ there is no trace of the initially dominant frequency left, and only a clean oscillation at 130.98 GHz, the previous sideband frequency.
This shows significant resemblance to the frequency jumping regularly observed in experiments, even though in simulation such jumps were the exception to the rule. On the other hand, the total time span in simulation for this behavior to occur and be observed is only 500 ns (i.e. the steady state phase), while the pulse lengths routinely monitored with the PSA system in experiments are by a factor of 10^4 larger than this value. Considering that in the experiment only very few jumps per pulse are found, it is clear that much longer simulation times would be required to get a statement on the incidence of the effect.

[27] With $\delta f \approx 10\,\mathrm{MHz}$, the spectra are only time-located within 100 ns.

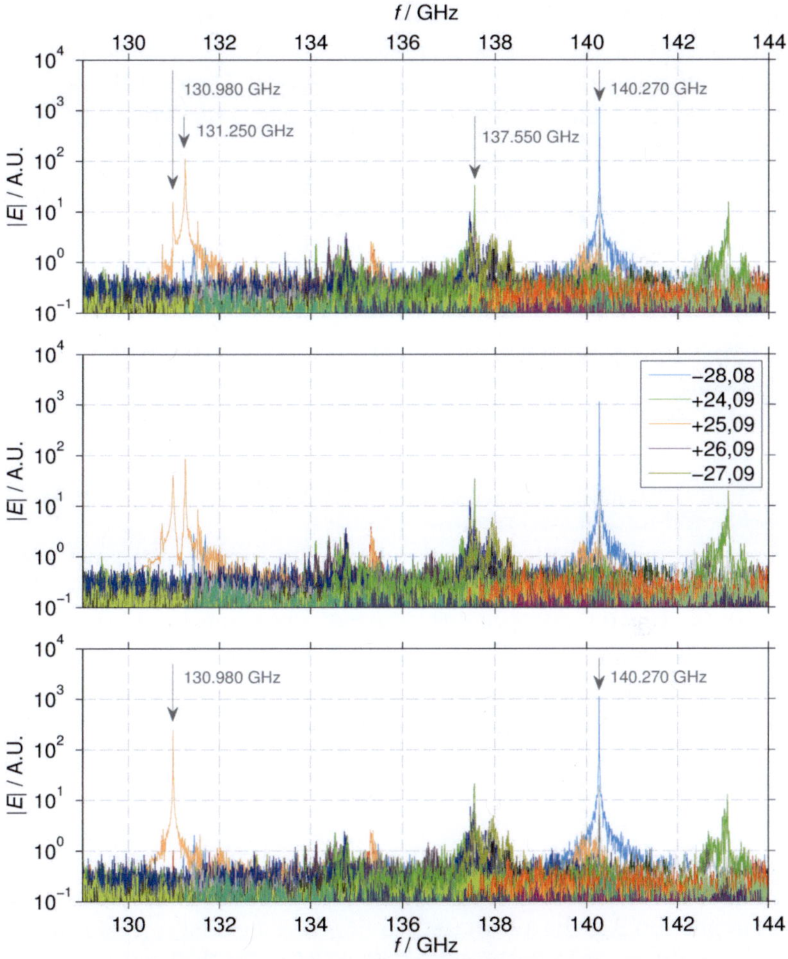

Figure 6.18: Amplitude spectra at indicated positions in Fig. 6.16
Top: 2200 ns, center: 2300 ns, bottom: 2500 ns

Other weak signals observed in the spectrum are around 137.55 GHz ($TE^{+}_{24,9}$ mode) and 143.095 GHz ($TE^{-}_{29,8}$ mode), to a certain extent symmetric to the main cavity mode. Although on rare occasions signals with quite low receiver power which could match these were measured, the relevance of these lines is uncertain. They could correspond to secondary excitations as they are described in [Nus81] as "phase-connected mode interaction".
The heterogeneous signals around 134 GHz are not sufficiently pronounced or reproducible to be interpreted.

6.4.2 Sweep over cathode voltage

In a first study, the entire measured cathode voltage sweep discussed in Section 6.2.2 is reproduced in simulation. Each measured voltage point is covered with an EVR-BZ start-up simulation as described in Section 3.2.2, and investigated for the time evolution and final steady state. Subsequent EVR-PIC continuation runs of 300 ns length were performed for selected operating points.
The results are compiled in Figure 6.19, where the measurement results known from Figure 6.4 are repeated. In addition to the frequencies found by EVR-BZ and EVR-PIC, the beam voltages for the cavity mode transition in measurement and simulation are also indicated. The parasites are coded with respect to the simulation tool and their relevance (major/minor), in a distinction similar to the one made for the representations in Section 6.2.2. Additional markers indicate the few cases of dynamic ACI, which were also found in the simulations.

Observations

As in Section 3.3, the simulated cavity mode transition takes place at higher voltages than in the experiment: While in the measurement, the first point without $TE^{-}_{28,8}$ mode dominance is at $V_b = 78.8\,kV$, in the simulations, the nominal cavity mode dominates up to $V_b = 82.2\,kV$. Except for this range of cavity mode disagreement, the simulated frequencies match the measurement better than ± 20 MHz.

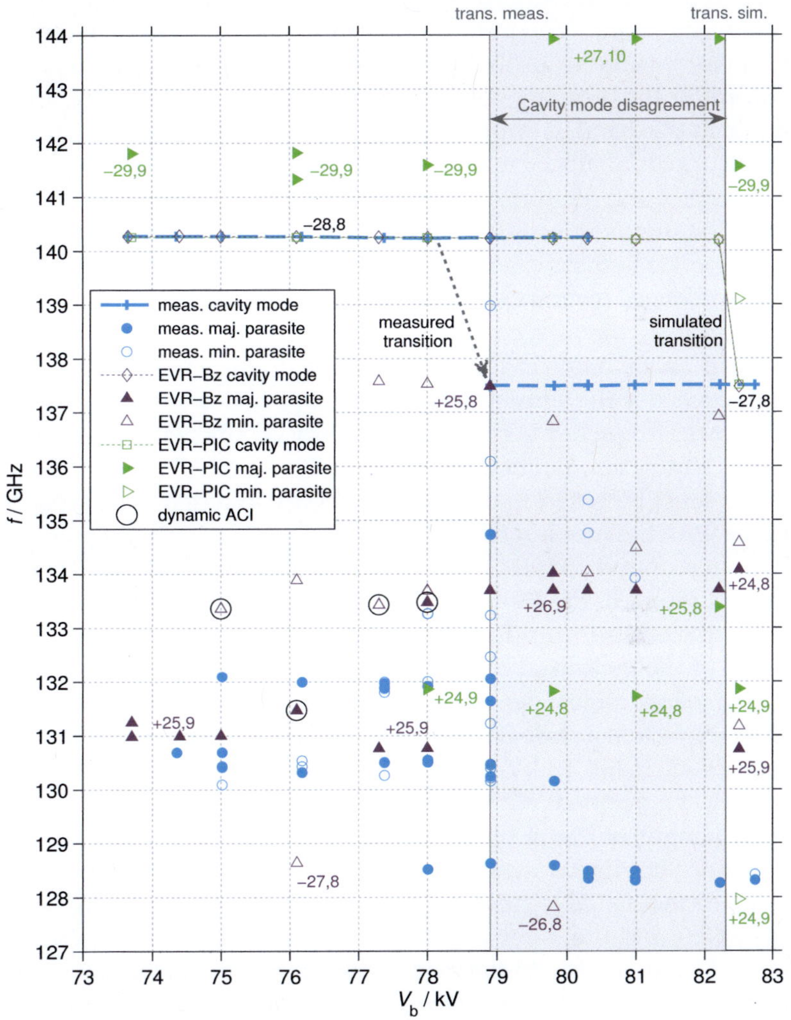

Figure 6.19: Measured and simulated frequencies for the V_{cath} sweep presented in Section 6.2.2

Concerning the EVR-BZ parasite frequencies, there is reasonable agreement with the measurement, especially concerning the formation of frequency bands. The parasite onset for low voltages is in the same frequency range as in the measurement, and shows similar tuning. Except for the outlying result with dynamic ACI at $V_\mathrm{b} = 76.2\,\mathrm{kV}$, the mode in this band at approximately 131 GHz is $\mathrm{TE}^+_{25,9}$.
In the upper frequency range 133–138 GHz, the are very few measured frequencies and these are scattered, which makes comparison problematic. The simulation suggests here considerably more activity, but still corresponds to the measured frequencies to a certain extent. While for lower voltages several dynamic ACI are predicted, strong general ACI of the $\mathrm{TE}^+_{26,9}$ mode occurs for higher voltages. In the measurement, another cavity mode is already operating in this voltage range, so the significance of the simulations in this range is doubtful.
The lower frequency band below 130 GHz is only represented by two minor simulation frequencies with the $\mathrm{TE}^-_{27,8}$ and $\mathrm{TE}^-_{26,8}$ modes, but still it must be noted that they are roughly in a matching frequency range with the measured band.
Parasites around 132 GHz, which were prominent in the measurement, are not found in the EVR-BZ results. Peculiarly, they are the only relevant parasites in the EVR-PIC continuation runs in the frequency range of the measured bands. They match the experimentally determined values almost perfectly; however, their onset is at higher values of V_b. Note that a shift along the V_b axis by the difference of the cavity mode switch, i.e. 3.4 kV, would move these points to lower voltages, in perfect agreement with the measured band around 132 GHz.
Surprisingly, at many beam voltages, the PIC simulation delivered parasitic activity far away from the measurement. Most prominent were the modes $\mathrm{TE}^-_{29,9}$ and $\mathrm{TE}^-_{28,9}$, the former as a general ACI around 141.8 GHz and the latter appearing frequency-locked[28] to the cavity mode at 140.27 GHz. In the many measurements made, such frequencies occurred only very seldom and at unstable operating points.[29]

[28] Because of this, it was not plotted in the graph for clarity.

[29] There is some similarity to the TCV 118 GHz gyrotron (see Section 5.2.1): In it, only parasites above the cavity mode's frequency, at 119 and 120 GHz, were measured.

Discussion

The explicit disagreement between the two simulation code variants concerning the PO frequencies, of course, lowers the overall significance of the results. Further, when comparing fixed V_b points, there is little agreement between the measured and simulated frequencies, as the mode activity ranges in the simulation are shifted with respect to the experimental ones. While this is very likely a modeling problem (compare the discussion in Section 3.3.7), the discrepancy between the two code variants is harder to explain. One possibility is that the differences in the implementation of EVR-PIC, especially the treatment of the interplay between the RF field update and the electron propagation, tends to promote different interactions than EVR-BZ; the introduced skew in the mode competition would then lead to a discrepancy.
A number of relevant agreements between simulation and measurement remain:

- The general band structure in matching frequency ranges is reproduced: Most of the relevant simulated PO frequencies lie very close to the measured bands. Especially for the two major parasite bands which appear with the nominal cavity mode, good correspondents can be found in the simulations.
- As far as discernible, the tuning slope $\Delta V_b / \Delta f$ within the bands also matches, for example compare the two $TE^{+}_{25,9}$ groups in the 73.5–78 kV part of the EVR-BZ simulation with the measurement.
- The $TE_{25,9}$ mode is the most relevant parasite in the frequency range around 131 GHz according to EVR-BZ, in substantial agreement with the Brillouin diagram analysis result (Section 6.3.6).

This indicates that the simulated phenomena are following the same physical rules and limitations as the measured ones. It is obvious that if, instead of a systematic parameter sweep, only one or a few unconnected operating points would have been simulated, wildly different conclusions would have been drawn, depending on the selection of beam voltage.

6.4.3 Sweep over beam current

Since the parasite frequencies do not seem to be reproduced in simulation on a point-to-point basis, rather the general behavior is of more interest. Because of this, in the following, a sweep with EVR-BZ and EVR-PIC over the beam current is made for the operating point $I_{\mathrm{m}} = 84.2\,\mathrm{A}$, $I_{\mathrm{tr}} = 3.75\,\mathrm{A}$, $V_{\mathrm{cath}} = 82\,\mathrm{kV}$, which is part of the Ω_0 sweep presented in Section 6.2.3. As the starting currents were investigated in this experiment, the parasite onset current in the simulation is an important indicator of the possibilities to predict parasites in the gyrotron design phase.
In Figure 6.20 the result of this simulation study is summarized. As in the previous study, the simulated major and minor parasites are marked, and the occurrence of dynamic ACI is indicated separately. All parasite frequencies measured during dominance of the $\mathrm{TE}^{-}_{28,8}$ mode in the cavity are plotted here as a function of their experimentally determined starting current (compare Fig. 6.5). Note that for these points, the marker filling separates between the frequencies found at $I_{\mathrm{m}} = 84.2\,\mathrm{A}$ and at other main coil current values. The EVR-PIC results are not plotted, as for all operating points only significant parasites in the range 141.38–141.79 GHz appeared, with the $\mathrm{TE}^{-}_{29,9}$ mode.
The predominant parasite in the EVR-BZ simulations appeared as the $\mathrm{TE}^{+}_{25,9}$ mode; in the diagram all points not labeled with a mode name or marked as ACI are for this mode.

Observations

In the experiment, the general onset of parasitic activity was found not far below $I_{\mathrm{b}} = 40\,\mathrm{A}$, while at the operating point used in the simulation the first parasite was measured at $I_{\mathrm{b,start}} = 42\,\mathrm{A}$ with 132.63 GHz. The simulation sweep, however, also finds parasites for lower beam current. Even at exceptionally low-current test points at $I_{\mathrm{b}} = 26\,\mathrm{A}$ and 30 A (not plotted), parasites are still oscillating.[30]

[30]For EVR-BZ, mostly dynamic ACI at 130 GHz and the $\mathrm{TE}^{+}_{25,9}$ mode at 131 GHz, and for EVR-PIC the notorious $\mathrm{TE}^{-}_{29,9}$ mode at 141.28 and 141.78 GHz were found.

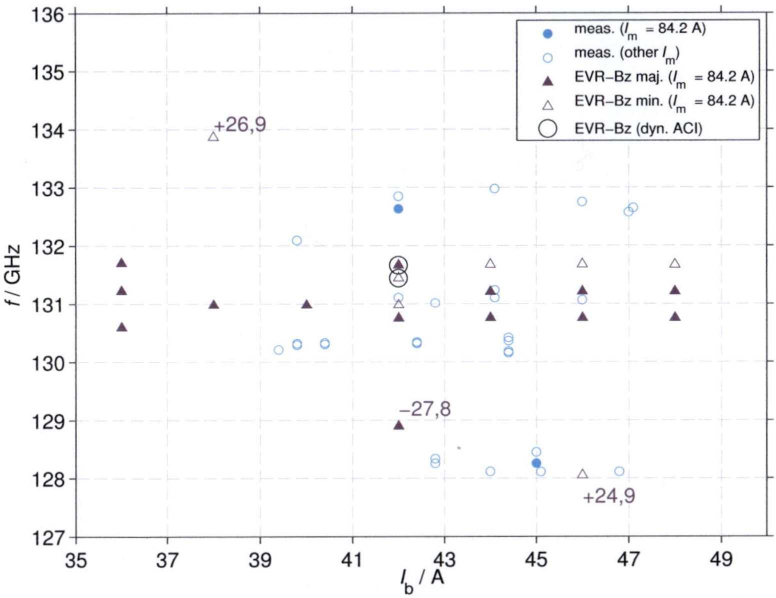

Figure 6.20: Parasitic frequencies of simulated beam current sweep for the $I_\mathrm{m} = 84.2\,\mathrm{A}$ point of the experimental I_m sweep (compare Section 6.2.3 and Fig. 6.5)

The major parasites in the EVR-BZ simulation are found around the frequency 131 GHz, where multiple peaks of the same mode, i.e. again the $\mathrm{TE}^+_{25,9}$ mode, are often present. This basically coincides with the two middle bands found in the measurements, in the range 130–131.4 GHz.

Except for this reasonable match, the other frequencies of the simulation appear scattered and unconnected. Few cases of dynamic ACI occur, however, without showing systematic behavior or notable matching with the experiment. Especially no frequencies in the range 132–133 GHz appear in the simulation, in contrast to the measurement. Two of the unconnected simulated frequencies match the lower band of measurements, 128–128.5 GHz, approximately in frequency and starting current. Unfortunately, this match is only found for two isolated points in different modes, i.e. $TE^{-}_{27,8}$ and $TE^{+}_{24,9}$, without much regularity. Also, in between the two points, at $I_b = 44$ A, in this frequency range no parasite appears in the simulation.

Discussion

The agreement between simulation and experiment in this sweep is not very satisfying, especially concerning the parasite starting currents. Since the simulation finds parasites even at I_b values more than 10 A below the experimentally determined starting currents, the possibility to reliably assess the danger of parasitic oscillations in a design phase is practically not given.
With EVR-BZ, a certain agreement with the measured frequencies was achieved. However, for example the band around 132.8 GHz, which was found around $I_m = 84.2$ A in the experiment, is missed entirely. Generous interpretation of the concurrent frequencies 130.75 GHz and 131.21 GHz above $I_b = 42$ A could make one think of tentatively matching the lower frequencies with the measured band of 130.3 GHz, but a comparison with Figure 6.5 reveals that these two bands never appeared simultaneously, reducing the justification of this view.
EVR-PIC is in principle better suited to treat after-cavity oscillation problems, as explained in Chapter 3. However, the simulations with this tool disagree to an even bigger extent with the experimental results, and raise more questions than they answer.

6.5 Summary

The opening classification for parasitic RF oscillations already indicated that from an experimental point of view, a basic important step is to locate the undesired interaction geometrically. Only after this, the more detailed investigation of the interacting mode and the possible reasons for its excitation in simulation are reasonable.

A number of approaches to this problem were introduced and combined, spanning in the measurement domain from detailed analysis of wide-band spectrograms gathered with the PSA system (see also Chapter 4) over the investigation of various parameter dependences in systematic experimental sweeps to a macroscopic overview of all measured frequencies. The combined interplay of the indications yielded by the different approaches is illustrated in Figure 6.21.

It was deduced from the spectral appearance that the parasites are probably forward waves, and both intra-pulse behavior and observations made in parameter sweeps indicate that the parasites are downstream of the cavity, i.e. after-cavity interactions (ACI). Especially the reproducible coincidence of the parasite emergence with the start of quasi-stationary cavity mode operation discussed in Section 6.2.1 would be hard to explain for BCIs, while it would represent natural behavior for an ACI.[31]

[31] A wide span of V_{cath}, I_{b} and (implicitly) Ω_{r} values was covered, as shown in Figure 6.6: In not a single measurement a parasite was detected before the cavity mode had assumed quasi-stationarity.

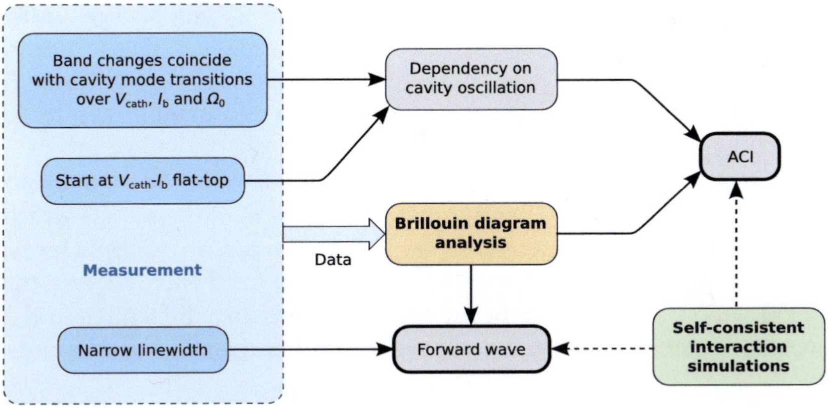

Figure 6.21: Overview of gathered indicators on parasitic oscillation properties

The analysis was continued in a fusion of experimental data with theoretical relationships, using the Brillouin diagram to estimate probable interaction positions and waveguide modes likely generating the observed signals. Although the method can only have limited precision due to the used uncoupled dispersion relation, and the discrete frequency jumping which seems to be characteristic of the current W7-X gyrotron parasites provides further interference, good matches were found for several batches of data. In all successful instances, forward wave ACI in the corrugated area of the launcher was diagnosed, with the $TE_{28,8}$ and $TE_{25,9}$ modes being prominent.
As a consequence, interaction simulations involving the trajectory code EVRIDIKI as well as its PIC variant were conducted, retracing a selection of the sweep measurements made. In a large number of simulations with an exceptionally extensive mode spectrum considered, several important principal agreements with the measured frequencies were found. Especially the general band formation and the reproduction of parasites in the range 130–131 GHz ($TE^{+}_{25,9}$), around 132 GHz ($TE^{+}_{24,8}$, $TE^{+}_{24,9}$) and several instances of dynamic ACI are to be noted here. However, important disagreements also emerged: The cavity mode transition was not properly reproduced, the simulated parasite starting currents are at least 10 A lower than measured, and generally more parasitic activity was predicted in the simulation. This indicates that with the currently available tools, interaction simulation alone is not capable of producing qualified statements on parasitic activity in high power gyrotrons.[32]
Although each of the individual methods applied has its own limitations and shortcomings, the significant agreement of the results which was yielded quite independently by all approaches on multiple levels is remarkable. It leads to the conclusion that forward wave ACIs in the corrugated part of the launcher are most probably the source of the weak parasitic activity in the current version of the W7-X gyrotrons.

[32]Note the discussion of possible sources for measurement-simulation discrepancies in Section 3.3.7.

7 Electrostatic and thermal effects during long pulses

The behavior of gyrotrons at the start of long pulses was introduced in Section 2.4.3, and illustrated by measurement in Section 5.1.2. In this Chapter, a more in-depth characterization in order to separate the underlying effects for their time constants and relevance is attempted.[1]

7.1 Overview

7.1.1 Relevance to gyrotron operation

The frequency shift caused by thermal expansion and space charge neutralization reflects a change of the operating point. Beam parameters gained from electron-optical simulations may thus only reflect the situation at the start of the pulse, while simulating the long pulse steady state requires assumptions about the final cavity radius expansion, and about the effective beam voltage as well. Typical approaches involve tuning the various parameters until a "good" match with the measurement results is achieved. While this might be acceptable for a-posteriori simulations, in a design phase without experimental data available this introduces a dangerous arbitrariness.
Simulating the transient phase between SP and stationary LP operation proves to be even more complicated, due to the unknown individual development of the two nonstationary processes. Mode losses or the rise of parasites in the time 10–1000 ms thus can only be treated experimentally, by empirical parameter adaption, which is not very satisfying.

[1] This involves principles from vacuum technology, which are compiled in the appendix Section C.

The trend in ECRH systems toward higher powers and frequencies [JAF+13] will also entail more strict electrical and mechanical tolerances for future gyrotrons. This increases the need for precise knowledge of the RF-relevant conditions inside the tube, in order for developers to ensure their designated area of operation not be too narrow. Advanced operation schemes of gyrotrons require pulse-modulated RF power generation for the goal of plasma stabilization [ZGG+01]. If the modulation pulse lengths happen to be of the same order of magnitude as the characteristic time constants of the tuning effects described here, it is desirable to be able to predict the exact behavior of the gyrotron in this regime. The effect of modulated operation on the gyrotron frequency was documented in [WGL+08].

7.1.2 Investigation approaches

The fact of frequency tuning at the start of longer pulses is well known, and nearly all experimental publications concerning long pulse operation note a frequency down-tuning of the cavity mode (e.g. [DBI+96, KST+01, BFB+04, DAA+06]). However, with the exception of [DBI+96], the effect is often presented as a matter of fact without further investigation, sometimes even neglecting the neutralization. As introduced in Section 2.4.3, the two principal tuning effects have independent physical sources, but influence the gyrotron interaction operating point and the frequency in superposition. Two basic paths of investigation are available:

- **Forward modeling**

  The processes are modeled as precisely as possible depending on the known physical quantities. This corresponds to simulations, which, if their result yields experimentally observable quantities, can be verified.

  Such an approach requires accurate modeling of both the thermal expansion process and the neutralization effect. Both of these phenomena have found some interest in the past; recently, thermo-mechanical simulations with COMSOL were conducted in [Cor13]. Problems arise from the values of material properties, such as the heat transfer coefficients or the electrical conductivity of the cavity wall at the operation frequency, which defines the thermal loss power. The modeling of the advanced cooling systems in the W7-X gyrotrons, including "Raschig-ring" structures [DAA+02], faces further difficulties.

The neutralization effects have been the subject of theoretical investigations on voltage depression to a certain extent in the past ([Pio90a, Pio90b, ZZM00, ZSST06]); however, no solid connections to the experimental situation are drawn, and in particular the gas dynamics are not given any consideration.[2] Forward modeling is employed in LP gyrotron development; in [AAB+01], the transient processes are explored with single mode interaction simulations to assess the danger of mode loss.

- **Backward analysis**

  This approach takes the opposite direction, as here the basis is the experimental data. After fitting, the physical processes are characterized by the parameters of modeling functions. The investigations made in [DBI+96] for a 0.5 kW 140 GHz gyrotron follow this track, focusing on the thermal cavity deformations; however, neutralization is also covered.[3]

  An advantage of this approach is the solid foundation on direct measurements, while the challenges are found in the formulation of an appropriate fitting configuration without too many degrees of freedom. This requires basic information on the relevant physical quantities, such as probable time dependence functions, so

[2] [Pio90b] contains the most systematic and straightforward approach for gyrotrons with axial output coupling, but no connection to the frequency tuning is made. The focus is on the effect on the beam parameters and on the pressure inside the tube.

[3] In [DBI+96], the primary source of experimental data for the various fits are various kinds of power measurements. The measured f-vs.-t information (see Fig. 6 ibid.) is not suitable for direct use, due to the lacking data quality in the first 10 ms of a pulse. Also, no advanced electron optics or interaction simulation codes, as they are used within this work, were available at the time.

that a certain amount of "forward modeling" is still required to establish the process boundary conditions. Despite its complementarity to theoretical methods and the inherently strong relation to experimental results, this approach has only seldom been pursued in the past.

The mentioned thermo-mechanical studies for W7-X gyrotrons suffered from uncertainties in the material constants, which increases the importance of experimental verification.
Because of this, after an overview of the involved processes, in the following the focus is put on the neutralization effects inside W7-X gyrotrons. After the establishment of an approximate macroscopic description from theoretical and technical relationships, the described "backward modeling" approach is taken, partly using and partly verifying the conclusions from the forward modeling process.
The main parameter for this investigation is the frequency of the cavity interaction, measured over time with the PSA system (compare Chapter 4) and its acquisition time expansion (see Section 4.7).

7.1.3 Connection of physical processes

A schematic diagram of the basic quantities and process dependences is given in Figure 7.1. Static influences, such as the magnetic field from the external gyrotron magnet, are not included. Of the beam parameters, only $W_\mathrm{k} = eV_\mathrm{b}$ is involved symbolically[4], and I_b.
Many important parameters cannot be measured experimentally with their time dependence, and even fewer of them can be influenced directly. To illustrate the inherent limitations to the experiment, the parameters in the diagram which can be set externally are marked by a thick border, while those which can be monitored over time have a thin double border.

[4]The pitch factor was omitted in the graph, but of course considered in all analyses. It is directly affected by V_dep, so it has the same process connections as W_k in Figure 7.1. It should be noted that this connection was neglected in [DBI+96], where the entire shift in instantaneous power was attributed to a change in V_b.

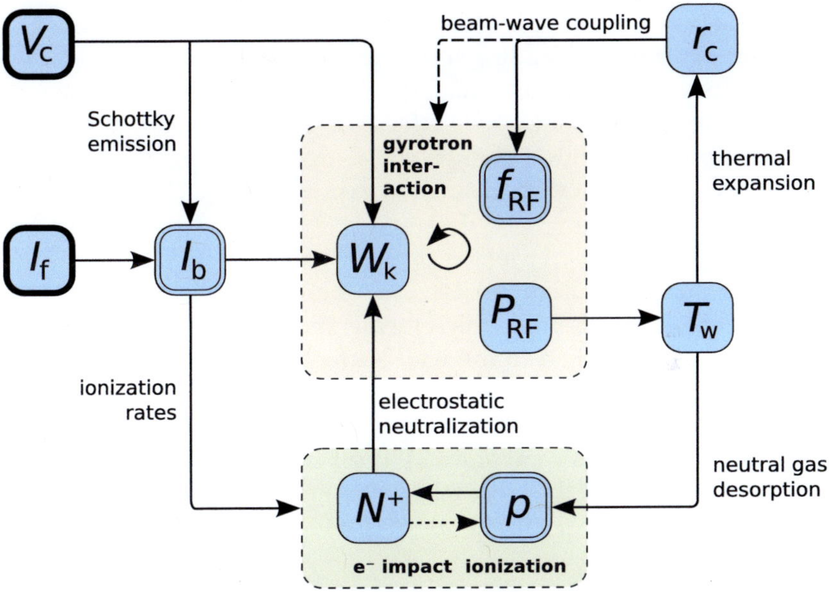

Figure 7.1: Basic parameter connections during longer pulses;
thick border: directly settable parameter
double border: time-dependent measurement available
single border: not settable, and no direct measurement

Connection descriptions

The beam current I_b is indirectly set by the filament heater current I_f, and follows V_{cath} (V_c in the graph) through the Schottky emission law, see Section 2.4.

In a first order approximation, the gyrotron interaction can be assumed to depend on W_k, f_{RF} and P_{RF}, linking these parameters in nonlinear feedback. For every change of one of these three parameters, a new operating point (W_k, f_{RF}, P_{RF}) will be established. As described in Section 2.1.3, $V_b = W_k/e$ differs from V_{cath} by the local electrostatic depression voltage V_{dep} at the start of the pulse.

From the produced RF power P_{RF}, wall losses arise which alter the wall temperature T_{w}, leading to thermal expansion, and therefore a change in the cavity radius r_{c}. Apart from the direct influence on the oscillation frequency, a side effect is the possible alteration of the output power through a change of the operating point. The influence on V_{dep} by this radius change as expressed in (2.12), however, is negligible.
Another consequence of wall temperature variation is a related change of the barometric pressure inside the tube. The main reason for this is the effect of outgassing, which means the temperature dependent release of gas adsorbed to the wall surface or absorbed to the bulk wall material itself.[5] An additional possible reason is the formation of micro-leaks at component boundaries under thermal deformation.
Both the rising wall temperature T_{w} itself and the resulting release of gas particles increase the barometric pressure inside the tube, which is obvious from the ideal gas equation[6] [Jou08]

$$\begin{aligned} p &= n\, k_{\mathrm{B}} T \\ \text{with} \quad p&: \text{pressure in Pa} \\ n&: \text{particle density in } 1/\mathrm{m}^3 \\ k_{\mathrm{B}}&: \text{Boltzmann constant, } 1.3806488 \cdot 10^{-23}\ \mathrm{J/K} \\ T&: \text{physical temperature in K} \quad . \end{aligned} \tag{7.1}$$

Together with the absolute value of the electron beam current I_{b}, the pressure governs the rates of the impact ionization process. This causes an effect of p and I_{b} on V_{b}, and therefore leads to a feedback connection between the neutralization and the thermal expansion. The superposition of the two effects is therefore not strictly linear.
Furthermore, the electron beam hitting the collector surface constitutes a second strong heating/outgassing source, which will partly shadow

[5] Compare Appendix C.4 for short descriptions of these effects.

[6] This equation describes the situation in low-pressure environments very well. In Appendix C.1, more information about the applicability of this equation for the conditions in gyrotrons is provided. Also note that in this work, the non-SI pressure unit mbar is used instead of Pa, for better compatibility with other publications.

the dependence of p on P_{RF}. The influences of these two effects could be divided by a $t_{\mathrm{p}} > 1\,\mathrm{s}$ experiment with constant RF power and swept efficiency. The low-efficiency part of this experiment would, however, place high loads on the collector, making such an experiment rather undesirable from a technological perspective.

7.2 Electrostatic neutralization investigations

In the following, an elementary way of describing the evolution of compensating positive charges in the interaction area of the gyrotron is presented. After the statement of a fundamental governing equation, the quantities in this equation are determined and their influence on the solution of the base equation for the application to W7-X gyrotrons is discussed.

All important statements on vacuum technology are taken from [Jou08, JW13], unless indicated otherwise. The relevant applied relationships from this field are summarized in Appendix C for reference.

7.2.1 Basic ionization process

As soon as the gyrotron pulse is started, the residual neutral gas in the tube is continuously ionized by the electron beam. The positive ions created by this electron-impact ionization process constitute a charge density $n_{\mathrm{i}}(t)$, which, following [Rei08], can be basically described through

$$\frac{\partial n_i}{\partial t} = g_{\mathrm{i}} \cdot n_{\mathrm{g}} = n_{\mathrm{b}}\sigma_{\mathrm{i}}v_{\|} \cdot n_{\mathrm{g}}$$

$$\begin{aligned} \text{with} \quad & g_{\mathrm{i}}\text{: ionization rate} \\ & n_{\mathrm{g}}\text{: neutral particle density} \\ & n_{\mathrm{b}}\text{: beam particle density} \\ & \sigma_{\mathrm{i}}\text{: ionization cross section} \\ & v_{\|}\text{: axial electron velocity} \quad . \end{aligned} \tag{7.2}$$

In this formulation, the neutral gas as well as the current density are considered homogeneous in the volume of interest with the cross section A_w along the beam trajectory of the length l_b. With

$$\frac{N_\mathrm{b}}{t} = \frac{I_\mathrm{b}}{e} \tag{7.3}$$

for the number of beam particles passing through an arbitrary transverse plane, the beam particle density in this volume $A_\mathrm{w} l_\mathrm{b}$ is

$$n_\mathrm{b} = \frac{I_\mathrm{b}}{e v_\parallel A_\mathrm{w}} \quad . \tag{7.4}$$

To obtain $\partial n_i / \partial t$ from (7.2), first the constituents on the right side and the rest gas species have to be found. In the process, the numerical ranges for typical W7-X gyrotron long pulse operation at $500\,\mathrm{kW}$ will be assigned.

7.2.2 Neutral gas in the gyrotron

Residual gas

An overview on gyrotron vacuum technology is given in [Dam95]. While the gyrotron is not operated, it constitutes a vacuum vessel which is continuously pumped by several ion-getter pumps (IGP, compare Appendix C.6), maintaining a specific residual pressure. A fraction of gas will be stored as a physisorbed surface layer, and another fraction will be chemisorbed in the vessel structure material; both cannot be not pumped by the IGP. However, for the gas species determination these effects do not matter.
The following relevant residual gas species exist:

- Nitrogen is used as a technical protection gas [Dam95], and is therefore a probable species for the residual gas. It also constitutes 78% of the atmosphere, making it the dominant species in the case of relevant leakage from the outside.

- Water vapor is a typical residual gas [Jou08], but is removed very efficiently during bake-out: Here the tube is heated to about 300–550°C while connected to a turbo-molecular pump for hours or days. This process also reliably removes hydrocarbons, which can be present in the form of oil and lubricant remains.
- With a pumping speed coefficient approximately one order of magnitude below the coefficients of other atmospheric gases (see [Jou08, Var94] and Appendix C.6), noble gases are also a potential candidate. However, they are usually only considered relevant when used as a protection gas, and their partial pressure in the normal atmosphere is below 1%, which is why we discard this possibility.
- Hydrogen can accumulate in a leak-tight vacuum device which is not pumped over longer amounts of time (at least weeks), and become the dominant residual species. This happens in a permeation process, i.e. diffusion through the vessel wall itself, due to the small molecular size of hydrogen molecules. For the gyrotron, it is still not a relevant residual gas candidate, since H_2 is pumped very efficiently by IGPs (compare Appendix C.6), and permeation rates are extremely low for the typical vessel wall thickness.

In the light of the above facts, nitrogen can be assumed to be the most relevant rest gas constituent.

Outgassing during operation

At the start of the pulse, the inside walls of the gyrotron will be quickly heated by RF radiation and, in the case of the collector, the impacting electrons. This causes physisorbed particle layers to evaporate quickly, while the chemisorbed particles will be gradually released with increasing structure temperatures over time. It should also be taken into account that the outgassing response of the gyrotron can be very inhomogeneous, as the temperature profiles of the various parts depend on their exposure to the internal RF radiation.

Figure 7.2 shows typical pressure developments inside the tube in the first 40 s of three pulses, as they were measured with standard mount IGP at the mirrorbox.[7]

These pulses were produced toward the end of the factory acceptance test of W7-X series gyrotron SN2i, that is, with a well conditioned tube, as is reflected by the very low pressure levels achieved. It is also obvious that outgassing takes place at the beginning of the pulse which increases the pressure by up to an order of magnitude, however, with limited reproducibility. During longer observation times than the ones plotted here, a slow and steady rise of the displayed pressure over many minutes is typical, resulting from the outgassing of components heated more slowly than the dominant ones at the start of the pulse and the release of chemisorbed particles.

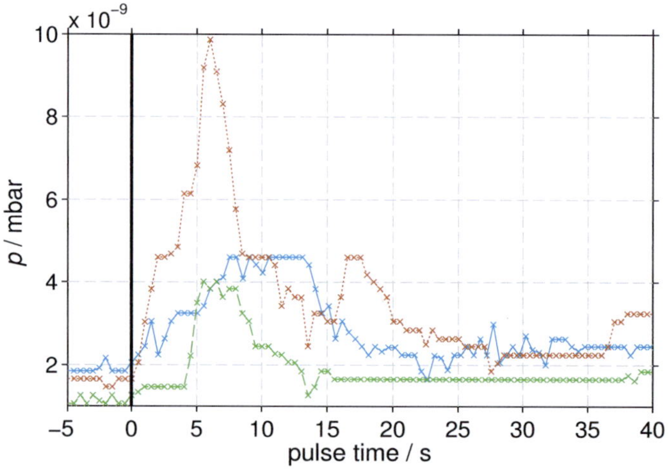

Figure 7.2: Measured vessel pressures during multiple 500 kW pulses with identical parameters (W7-X gyrotron SN2i)

[7] In the Appendix Sections C.5 and C.6, the basics and limitations for such measurements are established. The connection of these measured values with the true pressure and particle densities in the cavity area will be made later in this Chapter.

Note that the pressure measurement is very slow; here the acquisition is limited to two samples per second. Since the time constant of the processes investigated is less than one second, these data cannot be used directly to draw time dependence conclusions.

7.2.3 Ionization cross sections

In electron impact ionization processes, the cross section σ_{i} depends strongly on the kinetic energy of the electrons and on the ionization energy levels of the neutral gas. As a rigorous theoretical determination of σ_{i} involves complex quantum mechanical dependences which are problematic to solve even today [BS04], most practical approaches involve combinations of classical treatment with semi-empirical corrections from experiments. An overview of these techniques including comparison with experimental values is provided in [DBMM00].
Semi-empirical approaches yield considerable agreement with experimental results while retaining simplicity; however, most deal with low electron energies $W_{\mathrm{k}} < 10\,\mathrm{keV}$. An exception is the formalism by Slinker, Taylor and Rieke (condensed in [Rei08]), which estimates the ionization cross section as

$$\sigma_{\mathrm{i}} = \frac{8a_0^2 \pi R_{\infty} A_1}{m_{\mathrm{e}} c_0^2 \beta^2} f(\beta) \left(\ln \frac{2 A_2 m_{\mathrm{e}} c_0^2 \beta^2 \gamma^2}{R_{\infty}} - \beta^2 \right) \tag{7.5}$$

$$\text{with} \quad a_0\text{: Bohr radius, } 0.529 \cdot 10^{-10}\,\mathrm{m}$$
$$R_{\infty}\text{: Rydberg constant, } 13.6\,\mathrm{eV} \quad ,$$

which is parametrized for the gas species under consideration by the constants A_1 and A_2.
The correction function

$$f(\beta) = 1 - \frac{W_{\mathrm{i}}}{W_{\mathrm{k}}} \tag{7.6}$$

provides a measure for the relation between the gas ionization energy W_{i} and the kinetic energy of the electrons W_{k}, and is primarily relevant

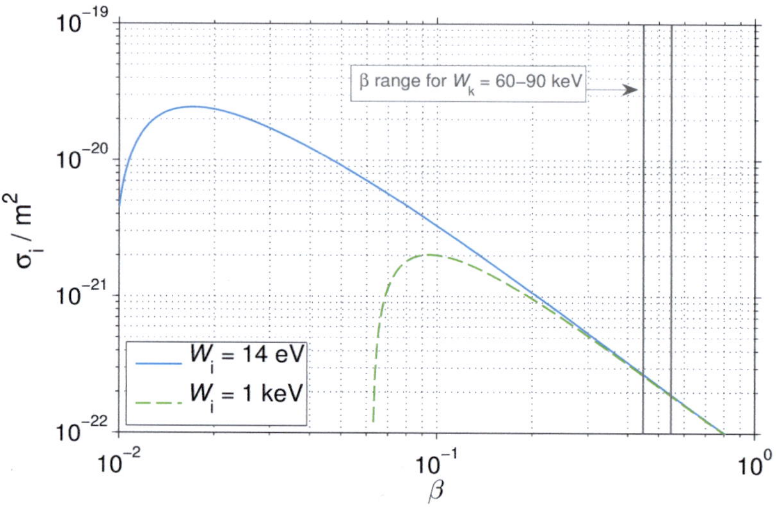

Figure 7.3: Cross sections σ_i for two different ionization energies W_{i} versus normalized electron velocity β

for electron energies in the order of the neutral gas ionization levels. For weakly relativistic electrons $f \to 1$, hence the ionization energy levels of the gas do not influence the cross section significantly.
This is obvious in Figure 7.3, where (7.5) was evaluated for two different ionization energies as a function of β. $W_{\mathrm{i}} \approx 14\,\mathrm{eV}$ is the first ionization level of N_2 (also of atomic N, compare e.g. [TK87]), while $W_{\mathrm{i}} = 1\,\mathrm{keV}$ is a conservative assumed upper limit for any ionization energy level relevant to this work.
For the electron kinetic energy range relevant in high power gyrotrons, $60 \lesssim W_{\mathrm{k}}/\mathrm{keV} \lesssim 90$, the two curves coincide, and energy-dependent values in the range $2.0 \cdot 10^{-22} \leq \sigma_{\mathrm{i}} \cdot \mathrm{m}^2 \leq 2.7 \cdot 10^{-22}$ are obtained.
The $W_{\mathrm{i}} = 1\,\mathrm{keV}$ curve also illustrates that there is only a relevant cross section for energies $W_{\mathrm{k}} \gtrsim W_{\mathrm{i}} := 1\,\mathrm{keV}$; this is intuitive, as the electrons with kinetic energies below this level have a negligible probability of causing an ionization event.

7.2.4 Ion confinement and energies

Only charged particles which are located in the z-range of the gyrotron interaction will contribute to the neutralization effect, and therefore have influence on the effective W_k. An unconfined ion will quickly hit the wall and re-neutralize by capturing an electron from the metal, hence returning to the fraction of neutral gas in the tube [Bit04].
Confinement can be achieved magneto- or electrostatically. It is worthwhile to also distinguish between confinement in the axial direction, and in the transverse dimension of the geometry.

Transverse confinement

At $t_p = 0$, the electron beam space charge creates a radial potential well of the depth Φ_{dep}, which is of the order of several kV (compare Section 2.1 and e.g. [PAI+14]). Positive ions with $W_{k,\perp} < \Phi_{dep}$ will inevitably be radially confined in this electrostatic well. While ions are accumulating and the potential is gradually neutralized, the confinement capability of the well is reduced.
However, even when the well is completely neutralized, or if high-energy ions are present, there remains magnetostatic confinement. Following (2.2), the ions will gyrate along the axial magnetic field lines. Transversal escape is only possible when their trajectory intersects the wall, i.e. $r_b + r_{L,ion} \gtrsim r_w$, assuming the generation of an ion in the vicinity of the beam guiding center r_b. Inserting the Larmor radius for a single charged[8] ion with the rest mass W_0, whose kinetic energy W_i is entirely available in the transverse direction, the $z-$local minimum kinetic energy for escape can be estimated to[9]

$$r_b + r_L = r_w \quad \Rightarrow \quad W_{i,escape}/eV = \frac{e\left[c_0 B_z \left(r_w - r_b\right)\right]^2}{2W_0} \quad . \qquad (7.7)$$

For a typical W7-X gyrotron operating point with $B = 5.6\ T$ in the cavity and N_2^+, this energy amounts to approximately $6\ \text{keV}$.

[8] Higher charge results in smaller r_L and therefore stronger magnetic confinement.
[9] Since the nitrogen molecule rest mass $W_0 \approxeq 26.1\ \text{GeV}$, this situation can be solved nonrelativistically.

The confining potential expressed by this minimum escape energy from magnetic confinement $W_{\mathrm{i,escape}}(z) = -\Phi_{\mathrm{mag}}$ is shown in Figure 7.4, where the wall and beam radius for the nominal W7-X gyrotron operating point are also indicated. It should be kept in mind that the transverse confinement at the start of the pulse is considerably stronger, as then also electrostatic transverse confinement is present; the electrostatic confinement potential Φ_{dep} is also plotted in the graph.

Axial confinement

According to (2.12), the electrostatic depression profile is dependent on the axial electron velocity and the ratio $r_{\mathrm{w}}/r_{\mathrm{b}}$, leading to a z-dependent potential well depth. In Figure 7.4, along with other data this profile is given for a typical operating point with depressed collector operation. The profile shows that only a fraction of the total beam trajectory contributes to ion generation which affects the cavity beam parameters: Confined ions will oscillate axially inside the potential well surrounding the cavity, while ions generated outside will be repelled axially.
As soon as the cavity potential well is compensated, the lack of axial confinement allows ions to follow the potential gradient towards the gun and leave the area of interest. In this case, the dominant motion is parallel to the magnetic flux lines, and no magnetic mirror effect can be expected since B_{z} decreases when moving away from the cavity in the z direction; thus no confinement prevents the ions from escaping along the axis. Over-neutralization of the beam is in principle possible, but restricted to quite specific cases.
Considering the assumption already stated that escaped ions will re-neutralize, as soon as the confining potential well is compensated, the local net ionization process ends ($\partial n_i/\partial t \approx 0$). Within the relevant geometry range, it is therefore useful to express the compensation status as a neutralization factor

$$\zeta = \frac{N_{\mathrm{i}}}{N_{\mathrm{b}}} = \frac{n_{\mathrm{i}}}{n_{\mathrm{b}}} = 1 - \frac{V_{\mathrm{dep}}}{V_{\mathrm{dep,0}}} \quad , \tag{7.8}$$

where $V_{\mathrm{dep,0}}$ denotes the voltage depression at the start of the pulse.

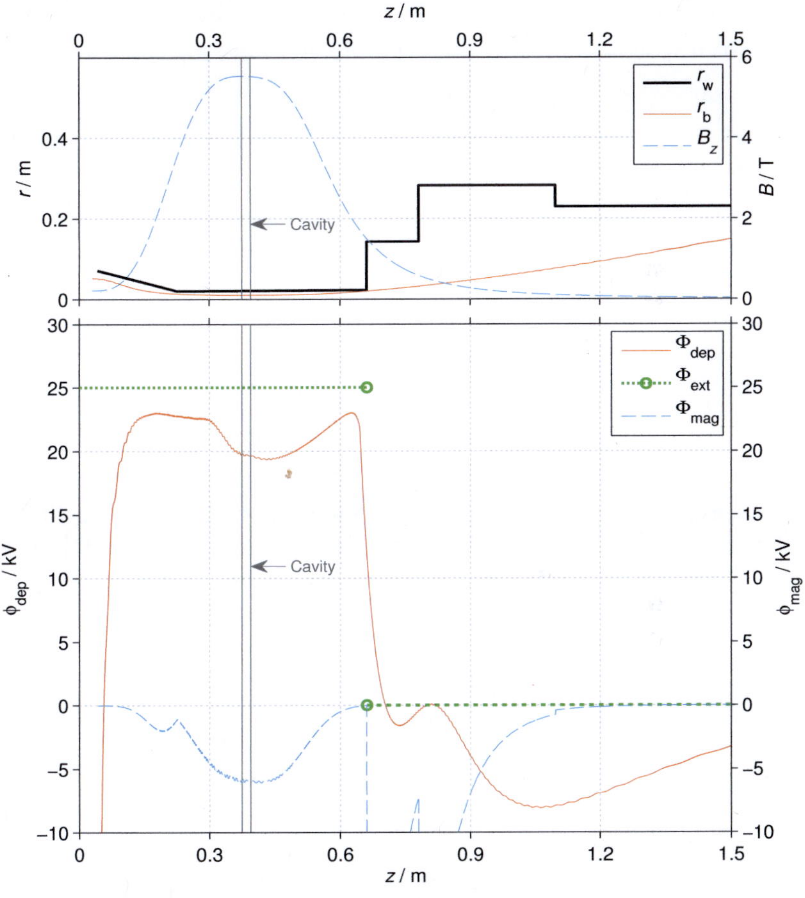

Figure 7.4: **Top:** Simplified circular waveguide geometry model of a W7-X gyrotron with B_z profile and electron beam radius **Bottom:** Axial and transversal confinement in sample W7-X gyrotron operation

Another important consequence of the axial potential profile in Figure 7.4 is the fact that at least for the W7-X gyrotron configuration, full neutralization ($\zeta \approx 1 \leftrightarrow V_{\mathrm{dep}} \approx 0$) cannot be achieved. The relevant potential well in $180 < z/\mathrm{mm} < 625$ is compensated when

$$\frac{N_{\mathrm{i}}}{N_{\mathrm{b}}} =: \zeta_{\mathrm{end}} \approxeq 0.6 \tag{7.9}$$

in this axial range, which will end net neutralization. As only the ionization process within the extend axial extent $l_{\mathrm{b}} \approx 445\,\mathrm{mm}$ of this cavity potential well (CPW) is relevant for the behavior of the cavity mode, it will be described by the geometrical properties $V_{\mathrm{cpw}} = A_{\mathrm{w}} l_{\mathrm{b}}$ as

$$\frac{\partial N_i(t)}{\partial t} = \begin{cases} g_{\mathrm{i}} V_{\mathrm{cpw}} n_{\mathrm{cav}}(t) & \text{if } N_{\mathrm{i}} < \zeta_{\mathrm{end}} N_{\mathrm{b}} \\ 0 & \text{if } N_{\mathrm{i}} \approx \zeta_{\mathrm{end}} N_{\mathrm{b}} \end{cases} . \tag{7.10}$$

Kinetic energy

Knowing the depth of the confining potential wells raises the question about the typical energy of the confined ions. The thermal energy which the gas particles can adapt from the vessel walls is roughly $1\,\mathrm{eV}$; considering the depth of the axial and magnetic confinement wells presented above, it is clear that this is many orders of magnitude too low for having any effect on confinement.

Colliding primary electrons, however, transfer part of their large kinetic energy to the neutral gas particles.[10] For the simple considerations made here, we employ the "hard-sphere" kinetic model as it is widely used for estimations in general gas ionization problems [Nas71] or plasma physics [Bit04]. A basic case to estimate the maximum expected energy transfer is that of a central collision between a low-mass incident particle with m_1 and v_1 with an initially resting heavy particle

[10]Note that this effect is in addition to the ionization process itself, which is a quantum-mechanical interaction of the beam electron with the electron shell of the gas molecule [Bit04], and is already covered by the ionization cross section formalism (7.5). The kinetic energy of the beam electrons is much larger than the $14.5\,\mathrm{eV}$ for the first ionization level of nitrogen, so their energy after an ionization event can be considered unaltered.

with m_2 and $v_2 = 0$ before the collision. From the laws of energy and momentum conservation, the velocity distribution v_1', v_2' after the collision can be analytically calculated.[11] Due to the strong mismatch of the involved masses, only a negligible maximum kinetic energy transfer of a few eV for the typical energies of 60–90 keV in a single collision can be expected.
We therefore assume all ions to be initially confined; however, for very long pulse durations a kinetic energy buildup through multiple collisions and thermalization is thinkable, facilitating gradual ion escape. For this to relevantly affect V_b, unrealistically low pressures must be present in the tube, as only then the escaped ions cannot be replaced by new ionization events.

Consequences

Before steady state is reached and $\zeta \approx \zeta_\mathrm{end}$, any ions generated in the range of the CPW can be considered confined, as the approximate ion energies are orders of magnitude below the confinement energies. This eases the calculations considerably.
As the relevant range in the axial dimension is now defined, the base ionization rate g_i can be calculated, using $v_\parallel$ and n_b. The average axial velocity in the CPW range for a typical W7-X gyrotron operating point of $V_\mathrm{cath} = 80\,\mathrm{kV}$ is equal to $\overline{\beta}_\parallel = 0.36$, and the average cross section area is $A_\mathrm{w} \approx 1.5 \cdot 10^{-3}\,\mathrm{m}^2$. With this, the equivalent beam particle density as defined in (7.4) can be calculated to $n_\mathrm{b} \approx 1.04 \cdot 10^{15}\,\mathrm{m}^{-3}$, or $N_\mathrm{b} \approx 6.95 \cdot 10^{11}$.
Together with the ionization cross section $\sigma_\mathrm{i} = 2.3 \cdot 10^{-22}\,\mathrm{m}^2$ and $I_\mathrm{b} = 27\,\mathrm{A}$ for operation at 500 kW, the base ionization rate is

$$g_\mathrm{i} = n_\mathrm{b}\sigma_\mathrm{i}v_\parallel = \frac{I_\mathrm{b}}{eA_\mathrm{w}}\sigma_\mathrm{i} \approx 26\,/\mathrm{s} \quad . \tag{7.11}$$

[11] This is a nonrelativistic approach. Relativistic calculations involving the momentum-energy four-vectors are required if the total energy of particle 1 is in the same order of magnitude as the total energy of particle 2. However, for N_2 molecules the rest energy of about $W_2 \approx 26.1\,\mathrm{GeV}$ by far exceeds the total energy of the beam electrons $W_1 = \gamma_1 m_{\mathrm{e},0} c_0^2 \lesssim 600\,\mathrm{keV}$, justifying the presented treatment of energy and momentum transfer.

7.2.5 Particle densities and flows

The neutralization time dependence $\zeta(t)$ is governed by $N_{\mathrm{i}}(t)$, as defined above in (7.8). From (7.10), it is clear that knowledge of $N_{\mathrm{cav}}(t)$ is necessary to establish $N_{\mathrm{i}}(t)$.
In gyrotrons, and generally in vacuum tubes, the gas pressure is desired to be minimal during operation to prevent irrevocable emitter material poisoning and to minimize deleterious collisions of electrons with neutral gas particles. In practice, the internal pressures range is in the ultra-high vacuum domain below $10^{-6}\,\mathrm{Pa} \mathrel{\widehat{=}} 10^{-8}\,\mathrm{mbar}$ [Dam95]. As a major consequence, particle interaction of the gas molecules can be entirely neglected, allowing the use of the ideal gas equation (7.1).[12]

Mirrorbox/collector particle density (steady state)

As described in Section C.6, the IGPs mounted on the W7-X gyrotron allow a relatively direct measurement of the neutral particle density in the gyrotron's mirrorbox.[13] For calculations, however, it is useful to convert these pressures into particle densities, to remove temperature dependences leading to uncertainty in the calculations.
It must be considered that the measured particle densities apply to the IGP's chamber. In order to estimate the values inside the mirrorbox, the conductivity C_{conn} of the IGP connector, which consists of a cylindrical section and a metallic grill for RF protection, must be taken into account. It can be analytically estimated (compare Appendix C.3) as $C_{\mathrm{conn}} \approx 110\,\mathrm{l/s}$, yielding with (C.10) an effective IGP flow rate of $S_{\mathrm{eff}} = 13.2\,\mathrm{l/s}$ per IGP.[14] With the specified basic pump flow rate S_{IGP}, this can also be expressed as a conversion factor between the measured particle density at the pump n_{IGP} and the particle density n_{vess} in the mirrorbox-collector vessel, as

$$\frac{n_{\mathrm{vess}}}{n_{\mathrm{IGP}}} = \frac{S_{\mathrm{IGP}}}{S_{\mathrm{eff}}} \approx 1.137 \quad . \tag{7.12}$$

[12] Compare also the discussion on the Knudsen number in Appendix C.
[13] The pressure developments given in Figure 7.2 were gathered from this source.
[14] The effect of thermal effluence, i.e. an increase of a component's flux conductivity with its temperature, was considered, and has no relevant effect.

Detailed time-dependent information as desired for the investigations pursued in this work is not available from the IGPs, mainly because of the coarse time resolutions they provide.[15] From the example pressure curves, however, it was established that in a well-conditioned gyrotron the mirrorbox particle density increases by less than an order of magnitude in the first few seconds of a pulse.

CPW gas particle density (steady state)

The neutralization process depends strongly on the particle density in the range of the cavity potential well. Before the start of the pulse, the background particle densities inside the gyrotron are defined by the conductivities between the different compartments, as already used above; only if no pump was active, would the interior pressure be homogeneous.
For a coarse and conservative estimate, the connection between mirrorbox and cavity range can be approximated by a cylindrical tube[16] with $l = 150\,\mathrm{mm}$ and $r = 22\,\mathrm{mm}$, which yields a conductivity of $C_\mathrm{t} = 40\,\mathrm{l/s}$ with the relationships discussed in Appendix C.3. The static gradient between the mirrorbox and the center of the CPW can thus be estimated analog to (7.12) as

$$\frac{n_\mathrm{cav}}{n_\mathrm{vess}} \leq 1 + \frac{4S_\mathrm{eff}}{C_\mathrm{t}} \approx 2.071 \quad . \tag{7.13}$$

Flow continuity

Changes in n_cav during a gyrotron pulse translate into a net particle flow $q_\mathrm{N}^\mathrm{cav}$, for which a conservation equation can be defined as

$$\frac{\partial N_\mathrm{cav}}{\partial t} = q_\mathrm{N}^\mathrm{cav} = q_\mathrm{N}^\mathrm{vess} + q_\mathrm{N}^\mathrm{w} + q_\mathrm{N}^\mathrm{i} \quad . \tag{7.14}$$

[15] Compare Section 7.2.2 and Fig. 7.2.

[16] The extent of the CPW complicates this definition, as the downstream end is connected quite directly to the mirrorbox, while the upstream end is at a distance of more than 350 mm. Here, an average value was chosen.

The following particle flows are involved:

- $q_{\mathrm{N}}^{\mathrm{vess}}$, net flow to or from the larger vessel consisting of the mirrorbox and the collector, depending on n_{cav} and n_{vess}.
- $q_{\mathrm{N}}^{\mathrm{w}} > 0$, expressing the net desorption of gas from the cavity wall.
- $q_{\mathrm{N}}^{\mathrm{i}} = -\partial N_{\mathrm{i}}/\partial t$, the transfer of gas particles to the fraction of ions by the ionization process expressed as a gas flow, with the limitation included in (7.10).

The outgassing process here provides the largest uncertainty, as only extremely indirect information about it is available. For the following first order investigations, it is therefore neglected.

7.2.6 Time development scenarios

Depending on the magnitudes of the flows defined in (7.14), the instantaneous absolute particle numbers/densities become relevant for the form of $\zeta(t)$. The starting value $N_{\mathrm{cav},0}$ is known from the background pressure measurement. Setting $q_{\mathrm{N}}^{\mathrm{w}} = 0$, two fundamental cases for the flow ratios can be separated in (7.14):

1. $|q_{\mathrm{N}}^{\mathrm{vess}}| \geq |q_{\mathrm{N}}^{\mathrm{i}}| \quad \Rightarrow \quad \partial N_{\mathrm{cav}}/\partial t \approx q_{\mathrm{N}}^{\mathrm{vess}}$

   For significant neutral gas flows, the particle density change in the cavity potential well is not influenced by ionization, so its time dependence can be obtained through integration:

$$N_{\mathrm{cav}}(t) = \int_0^t q_{\mathrm{N}}^{\mathrm{vess}}(\tau)\mathrm{d}\tau + N_{\mathrm{cav},0} \qquad (7.15)$$

2. $|q_{\mathrm{N}}^{\mathrm{vess}}| \ll |q_{\mathrm{N}}^{\mathrm{i}}|$

   When $q_{\mathrm{N}}^{\mathrm{cav}}$ is not dominated by the flows, the confinement capability of the potential well and the availability of neutral gas particles become relevant.[17]

[17] For $q_{\mathrm{N}}^{\mathrm{vess}} = 0$, this reduces to the "closed system" cases discussed in [Pio90a, Pio90b].

In 2., thus the following sub-cases are found:

a) $N_{\mathrm{cav}} \gg \zeta N_{\mathrm{b}} \quad \Rightarrow \quad N_{\mathrm{cav}} \approx \mathrm{const}$

The ionization process stops before the N_{cav} can be influenced relevantly; therefore in $t < \tau_{\mathrm{n}}$, the neutral gas particle number is time independent.

b) $N_{\mathrm{cav}} \approx \zeta N_{\mathrm{b}} \;\Rightarrow\; \partial N_{\mathrm{cav}}/\partial t \approx -g_{\mathrm{i}} N_{\mathrm{cav}} \;\Rightarrow\; N_{\mathrm{cav}}(t) = N_{\mathrm{cav},0} \cdot \mathrm{e}^{-g_{\mathrm{i}} t}$

The ionization process reduces the rest gas with g_{i}. This is expressed by a homogeneous first order differential equation, which yields an exponential reduction of the neutral gas fraction with the time constant $\tau_{\mathrm{i}} = 1/g_{\mathrm{i}}$.

c) $N_{\mathrm{cav}} < \zeta N_{\mathrm{b}}$

There is enough confinement capability to deplete the neutral gas in the relevant area. This case gives rise to subsequent ionization reactions, where first with $\mathrm{N}_2^+ \rightarrow \mathrm{N}_1^+ + \mathrm{N}_1^+$ the ionized nitrogen molecules are dissociated into ionized atomic gas particles, which are then again ionized:[18]

- $\dfrac{\partial N_{\mathrm{cav}}}{\partial t} = -\dfrac{N_{\mathrm{cav}}}{\tau_{\mathrm{i}}}$ - molecule ionization
- $\dfrac{\partial N_2^+}{\partial t} = \dfrac{N_{\mathrm{cav}}}{\tau_{\mathrm{i}}} - \dfrac{N_2^+}{\tau_{\mathrm{diss}}}$ - molecule dissociation
- $\dfrac{\partial N_1^+}{\partial t} = 2\dfrac{N_2^+}{\tau_{\mathrm{diss}}} - \dfrac{N_1^+}{\tau_1}$ - formation of 2+ ions from 1+ ions
- $\dfrac{\partial N_1^{2+}}{\partial t} = \dfrac{N_1^+}{\tau_{\mathrm{diss}}} - \dfrac{N_1^{2+}}{\tau_2}$ - formation of triple charged ions ...
- ...

This defines a cascade of exponential functions, each ionization cross section being smaller than the previous. The maximum amount of charge created by this process is $14e \cdot N_{\mathrm{cav},0}$.

[18] For the mono-atomic particles, τ_x is the time constant of $\mathrm{N}^{x+} \rightarrow \mathrm{N}^{(x+1)+}$.

The typical displayed pressure range 10^{-9}–10^{-8} mbar in experiment can be converted to a pressure density, to then be corrected with (7.12) for the expected conditions in the CPW area. With $C_{\rm t} = 0.04\,{\rm m}^3/{\rm s}$, this yields background flows of $q_{\rm N}^{\rm vess} = C_{\rm t} n_{\rm vess} \approx 1.1\text{-}11 \cdot 10^{12}\,/{\rm s}$ between the CPW and the mirrorbox.
An instantaneous ionization particle flow $q_{\rm N}^{\rm i}$ is given directly by the ionization rate defined in (7.11). Using $n_{\rm g} \approx 5.7 \cdot 10^{13}\,/{\rm m}^3$ for a displayed pressure of 10^{-9} mbar, this flow can be estimated to be $q_{\rm N}^{\rm i} = g_{\rm i} N_{\rm g} \approx 10^{12}\,/{\rm s} \approx q_{\rm N}^{\rm vess}$ in the CPW area.
The background flow is capable of replenishing the gas, so case 1. in the above list applies; because the ionization flow is quite similar to the neutral gas background flows, it is worthwhile to investigate also the absolute quantities. The result for $n_{\rm g}$ with the corrections in equations (7.12) and (7.13) was already given. With $n_{\rm b} \approx 10^{15}\,/{\rm m}^3$, it is clear that without the flow consideration, a depletion of the gas in the cavity would be expected, as happened for example in [Pio90b]. Note that since the volume of the mirrorbox-collector vessel $V_{\rm vess}$ is about three orders of magnitude larger than $V_{\rm cpw}$, it cannot be drained by the ionization/confinement process.
In a first order approach one can therefore assume $n_{\rm g} \approx {\rm const}$ and obtain $N_{\rm i}(t)$ by direct integration in (7.10). For a linear process, the time constant $\tau_{\rm n}$ is identical to the neutralization time $t_{\rm n}$, i.e. the time when $\zeta(t_{\rm n}) = \zeta_{\rm end}$, which for a displayed pressure of 10^{-9} mbar is

$$t_{\rm n} = \frac{\zeta_{\rm end} N_{\rm b}}{g_{\rm i} N_{\rm g}} \approx 400\,{\rm ms} \quad . \tag{7.16}$$

Note that the obtained result presented above has more the character of an upper limit, as local outgassing was neglected. It can thus be stated that at least for operation at reasonable pressure, there is no possibility for very slow neutralization, i,e. exceeding a few seconds to be complete, in modern high power gyrotrons.
Depending on the background vacuum conditions, respectively outgassing, the neutralization process can thus be expected to complete within some hundred milliseconds for good vacuum conditions, and possibly within a few milliseconds or less for "bad vacuum".

7.3 Time dependence fitting

In the following, the developed fitting method is described and applied to a sample operating point. The used time-dependent experimental data, i.e. f-vs.-t traces $f_{\mathrm{meas}}(t)$, were gathered with the undersampling technique described in Section 4.7.2 and applied in Section 5.1.2.

7.3.1 Operating point transition

The pulse starts with a fully-depressed operating point, being characterized by the beam voltage $V_{\mathrm{b},0} = V_{\mathrm{cath}} - V_{\mathrm{dep}}$ and the associated beam parameters. The start frequency f_0 is a well known quantity from measurement. From the considerations in Section 5.1.3 it is already known that the "real" steady state depends, among other things, on the cooling system. The tuning observed in the first two seconds, however, is of the order of hundreds of MHz, while during the rest of the pulse a detuning of only approximately 1 MHz/min is found. Hence, we here ignore this long-term tuning, and assume steady state at $t_{\mathrm{p}} \geq 2\,\mathrm{s}$.
The thermal cavity expansion starts at the known cold radius $r_{\mathrm{c},0}$ with the starting temperature $T_0 \approx 290\,\mathrm{K}$, and ends with the establishment of a stationary equilibrium between RF loss and cooling power, at T_{end} with the unknown $r_{\mathrm{c,end}}$. The absolute deformation[19] of the cavity at steady state is

$$\Delta r_{\mathrm{c}} = r_{\mathrm{c,end}} - r_{\mathrm{c},0} \quad . \tag{7.17}$$

From the initial value $r_{\mathrm{c}} = r_{\mathrm{c},0}$ to the steady state with $r_{\mathrm{c}} = r_{\mathrm{c,end}}$, a parameter trajectory is described by the deformation function $\Delta r_{\mathrm{c}}(t)$, which contains the form of the time dependence. Only $r_{\mathrm{c},0}$ is a fundamentally known quantity. Often experimentalists try to directly

[19] The form of the cavity deformation has already been studied in the past with thermo-mechanical simulations, e.g. in [Ber11, Cor13]. The results suggest that the cylindrical section expands non-uniformly into a "barrel-shape". However, in comparing interaction simulations, the resulting deviation between barrel-shaped and uniform expansion was negligible, which is why in the studies presented here simply equal expansion is assumed.

estimate the total radial expansion $\Delta r_{\mathrm{c,end}}$, directly using (5.1):

$$\frac{r_{\mathrm{c,end}}}{r_{\mathrm{c,0}}} = \frac{f_0}{f_{\mathrm{end}}} \quad \rightarrow \quad \Delta r_{\mathrm{c,end}} = \left(\frac{f_0}{f_{\mathrm{end}}} - 1\right) r_{\mathrm{c,0}} \quad . \qquad (7.18)$$

This entirely ignores the effect of the electrostatic neutralization, and is not applicable here.

As investigated in Section 7.2.4, the effective beam voltage at steady state does not necessarily equal V_{cath}. We allow for only partial neutralization by defining

$$\Delta V_{\mathrm{b,end}} = V_{\mathrm{b,end}} - V_{\mathrm{b,0}} \quad , \qquad (7.19)$$

and again a time dependence $\Delta V_{\mathrm{b}}(t)$ in analogy to (7.17). The connection to the previously introduced neutralization fraction ζ is made using (7.8), and the steady state values for frequency as well as the RF power, f_{end} and P_{end}, are obtained from the experiment.[20]

Since $\Delta V_{\mathrm{b}}(t)$ is the only transient quantity which influences the beam parameters, these can easily be simulated in the whole relevant $\Delta V_{\mathrm{b,end}}$ range $0\,\mathrm{kV}$–V_{dep} as a prerequisite. The typically linear behavior of beam parameters in such a small voltage range also makes it possible to reduce the beam parameter simulation to the extremal values.

7.3.2 Simulated parameter map

With the beam parameters available, a parameter study of instantaneous power and frequency as a function of the two variables can be conducted. For this, a grid of relevant $[\Delta r_{\mathrm{c}}, \Delta V_{\mathrm{b}}]$ combinations is defined and simulated in the interaction code EVRIDIKI.[21]

In Figure 7.5, such a parameter map is presented. The operating point treated here and in the following yielded $500\,\mathrm{kW}$ with a good reproducibility at $V_{\mathrm{cath}} = 53.8\,\mathrm{kV}, V_{\mathrm{body}} = 24.7\,\mathrm{kV}$ with a beam current

[20] The measured power will not be used in this investigation as an input variable.

[21] As only frequency data are of interest, one could also use the analytical techniques introduced in Section 3.2.2, which means considerably less effort than full simulations. The frequency accuracies found in the Chapter 3 benchmarks, however, show that full interaction codes should be used when possible.

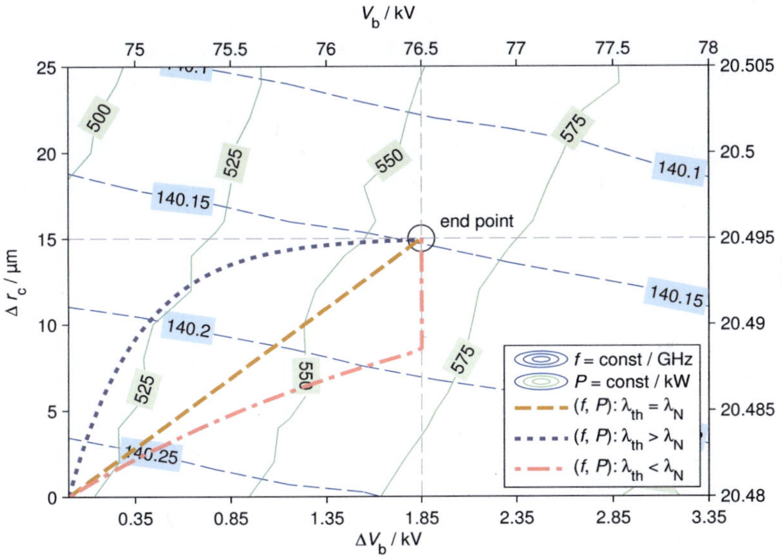

Figure 7.5: Simulated parameter map for swept cavity deformation Δr_{c} and effective beam voltage increment ΔV_{b} with three hypothetical temporal development curves

of $I_{\mathrm{b}} = 27.3\,\mathrm{A}$. An initial voltage depression of $V_{\mathrm{dep}} = 3.9\,\mathrm{kV}$ was estimated with ESRAY, hence $V_{\mathrm{b},0} = 74.6\,\mathrm{kV}$.
As in a Rieke diagram (see e.g. [BJ90]), the map consists of perpendicular constant-frequency and constant-power lines. The operating point at the start of the pulse, i.e. inherently a SP operating point with nominal cavity radius and full voltage depression, is situated at the point of origin $[\Delta r_{\mathrm{c}}\,/\,\mu\mathrm{m}, \Delta V_{\mathrm{b}}\,/\mathrm{kV}] = [0, 0]$.
A given value of $\Delta V_{\mathrm{b,end}}$ determines a line vertical to the ΔV_{b}-axis on which the transient process ends. Together with f_{end} from the measured frequency curve, this allows the location of the final operating point $[\Delta r_{\mathrm{c,end}}, \Delta V_{\mathrm{b,end}}]$ on this map. This yields the final radial cavity deformation and the simulated stationary power, and was marked in Figure 7.5 for illustration purposes at the end point $[15\,\mu\mathrm{m}, 1.85\,\mathrm{kV}]$.

In such a diagram, a known trajectory of the frequency tuning processes is represented as a curve between the origin and the end point. Three different fictional tuning trajectories were inserted in Figure 7.5.[22] If the time constants of both processes are identical, the resulting trajectory resembles a straight line between origin and end point (dashed-orange line). In the case of $\lambda_{\mathrm{th}} \gg \lambda_{\mathrm{n}}$, i.e. fast expansion, the dotted-blue trajectory is obtained. For fast neutralization, $\lambda_{\mathrm{th}} \ll \lambda_{\mathrm{n}}$, an interesting effect can be observed: Since the neutralization ends abruptly before the major fraction of expansion has taken place, an edge in the curve is created. Beyond this point, the frequency tuning is purely caused by radial expansion.

7.3.3 First-order approach without outgassing

Time dependences

Based on the previously outlined considerations, the radial expansion process can be modeled as a exponential development[23] between $r_{\mathrm{c},0}$ and $r_{\mathrm{c,end}}$, while for the neutralization process a linear change from $V_{\mathrm{b},0}$ to $V_{\mathrm{b,end}}$ is expected:

$$r_{\mathrm{c}}(t) = r_{\mathrm{c},0} + \overbrace{\Delta r_{\mathrm{c}}\left(1 - \mathrm{e}^{-\lambda_{\mathrm{th}} t}\right)}^{\Delta r_{\mathrm{c}}(t):=} \tag{7.20}$$

$$V_{\mathrm{b}}(t) = V_{\mathrm{b},0} + \underbrace{\Delta V_{\mathrm{b}} \cdot \begin{cases} \lambda_{\mathrm{n}} t & \text{, if } \lambda_{\mathrm{n}} t \leq 1 \\ 1 & \text{, if } \lambda_{\mathrm{n}} t > 1 \end{cases}}_{\Delta V_{\mathrm{b}}(t):=} \quad . \tag{7.21}$$

For a fixed set of end points $[\Delta r_{\mathrm{c,end}}, \Delta V_{\mathrm{b,end}}]$, the development rates $\lambda_x = 1/\tau_x$ entirely define the temporal development. A given set $[\lambda_{\mathrm{th}}, \lambda_{\mathrm{n}}]$ can thus be directly converted into a "stiff" $f_{\mathrm{sim}}(t)$-curve by

[22]Here for $\Delta V_{\mathrm{b}}(t)$ a linear development and for $\Delta r_{\mathrm{c}}(t)$ an exponential increase was assumed, in accordance with the prior investigations.

[23]See also the investigations made in [DBI+96, Cor13] in this matter.

insertion of the trajectory $[\Delta r_c(t), \Delta V_b(t)]$ into the parameter map in Figure 7.5, even though the map itself does not contain any time information. The result obtained this way can then be assessed by comparing $f_{sim}(t)$ with the measured and true frequency development $f_{meas}(t)$. This can be done graphically, using the instantaneous frequency difference $\Delta f = f_{sim} - f_{meas}$, or numerically over the root-mean-square (RMS) value ϵ calculated from $\Delta f(t)$.
The separation of time constants thus can be formulated as an optimization problem in the $[\lambda_{th}, \lambda_n]$-plane with the goal function $\epsilon \to \min$ and a parameter $\Delta V_{b,end}$.

Optimization overview

The achievable neutralization fraction $\zeta_{end} = \Delta V_{b,end}/V_{dep}$ was estimated from the axial confinement consideration in Section 7.2.4 to be in the range 50–60%, but can also be investigated separately by a parameter sweep of the optimization process $\min(\epsilon\{\lambda_{th}, \lambda_n\})$.
Figure 7.6 shows the obtained minimum ϵ for a sweep over the $\Delta V_{b,end}$ range, on a common axis with the neutralization percentage.[24] In the top part of the figure, the influence of the assumed final neutralization fraction ζ_{end} on the expected total deformation $\Delta r_{c,end}$ is visible. For $\zeta_{end} = 0$, the complete frequency tuning is attributed to thermal expansion, while increasing ζ_{end} naturally causes lower thermal deformation values.
In the top graph, the optimization results ϵ_{min} are also plotted. The achieved global minimum ϵ_{min} clearly supports the hypothesis that only fractional neutralization is achieved, and is located at 2.2 kV, approximately at $\zeta = 0.57$. This is in extremely good agreement with the confinement considerations of Section 7.2.4; however, note that the optimization minimum is not very sharp, so statements more accurate than ± 0.25 kV are not justified.

[24] $\Delta V_b = 0$ kV would be 0% final neutralization, while $\Delta V_b = V_{dep}$ means 100% neutralization. Values in the range $0.9 < \zeta \leq 1$ were omitted, since here stable operation was not possible not for all datapoints in the simulation, leading to numerical instabilities. Small ΔV_b values did not affect stability, but obviously cause the resulting λ_n to be not very meaningful.

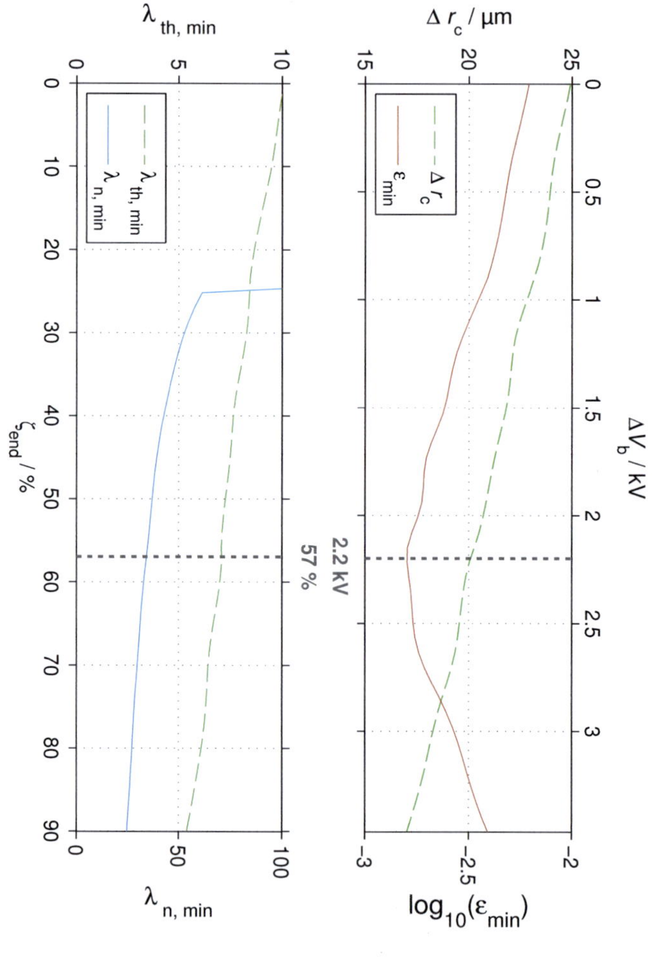

Figure 7.6: Dependence of the optimization results on the final neutralized depression voltage
Top: maximum Δr_c and minimum RMS residuals
Bottom: corresponding fit results

The bottom plot illustrates the related developments for the process rates $[\lambda_{\mathrm{th}}, \lambda_{\mathrm{n}}]$. Already without considering the minimum ΔV_{b}, a clear separation between the tuning process rates occurs, with $\lambda_{\mathrm{n}} > 25\,/\,\mathrm{s}$ and $5\,/\,\mathrm{s} < \lambda_{\mathrm{th}} < 10\,/\,\mathrm{s}$ indicating that the neutralization process is faster. At the minimum, the tuning rates $[\lambda_{\mathrm{th}}, \lambda_{\mathrm{n}}]\cdot\mathrm{s} = [6.9{-}7.3, 31{-}36]$, respectively the time constants $[\tau_{\mathrm{th}}, \tau_{\mathrm{n}}]/\mathrm{ms} \approx [137-145, 28-32]$ were obtained in such fits of multiple $500\,\mathrm{kW}$ pulses.

Process rate plane at optimum (RMS error map)

The quality of a fit result can be judged by the surroundings of the optimum in the parameter plane $\epsilon(\lambda_{\mathrm{th}}, \lambda_{\mathrm{n}})$; Figure 7.7 contains a "RMS map", constituted by the evaluation of $100 \cdot 100$ $[\lambda_{\mathrm{th}}, \lambda_{\mathrm{n}}]$-combinations for their respective RMS error value ϵ at $\Delta V_{\mathrm{b,end}} = 2.2\,\mathrm{kV}$.

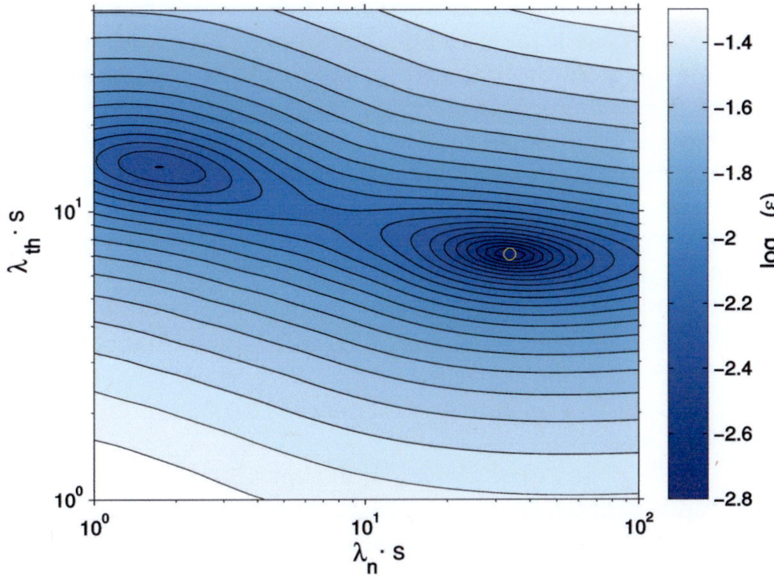

Figure 7.7: Map of $\log_{10}(\epsilon)$ in $[\lambda_{\mathrm{th}}, \lambda_{\mathrm{n}}] \cdot \mathrm{s} = [1 \ldots 40, 1 \ldots 100]$, with a small yellow circle marking the global ϵ minimum

An important feature of the RMS map is the contrast of the global minimum at $[\lambda_{\text{th}}, \lambda_{\text{n}}] \cdot s = [7.1, 34]$ (i.e. $[\tau_{\text{th}}, \tau_{\text{n}}]/\,\text{ms} \approx [141, 29]$), which defines the fit significance. While the contrast and definition of this minimum are good, a secondary minimum at $[\lambda_{\text{th}}, \lambda_{\text{n}}] \cdot \text{s} = [14.3, 1.7]$ is also discernible. It expresses a case of rather slow neutralization with $\tau_{\text{n}} = 590\,\text{ms}$, and faster expansion with $\tau_{\text{th}} \approx 70\,\text{ms}$. The relevance of this minimum is discussed in the next section.
The optimization overview in Fig. 7.6 contains the ϵ_{min} from such a global minimum in each point. It should be noted that while the optimization for a global minimum operates on a relatively flat curve, the minimum in each RMS map is very pronounced.

Frequency development and parameter map trajectories

A graphical comparison of $f_{\text{meas}}(t)$ and the fitting result for the global optimum is given in Figure 7.8, along with the instantaneous frequency difference $\Delta f(t)$ as a more detailed fit quality indicator. On both time scales the good fit match is clearly visible, with the maximum frequency deviation being well below $\pm 5\,\text{MHz}$.
With the found time constants, a valid trajectory can now be inserted in the parameter map for this operating point. In Figure 7.9, the same parameter map as previously shown in Figure 7.5 is given, but with the parameter trajectory corresponding to the global optimum from the fitting process.
The obtained stationary state power is about $565\,\text{kW}$; from long-pulse power balances as used in Section 3.1.3 it is known that 90% of the dissipated power in the cavity is absorbed in the main load. Considering this, the resulting operating point matches the measured power of $500\,\text{kW}$ very well.
Since $\lambda_{\text{n}} > \lambda_{\text{th}}$, the curve shows an abrupt kink at the end of the neutralization process, i.e. at $t_{\text{n}} = 1/\lambda_{\text{n}}$ as predicted previously for this case. Interestingly this now can be related to the $f(t)$ behavior shown in Figure 7.8, where $\dot{f}_{\text{meas}}(t)$ exhibits a discontinuity at $t = 30\,\text{ms}$ which is matched cleanly by $f_{\text{sim}}(t)$.[25]

[25] This phenomenon is only properly visible in the time-logarithmic representation.

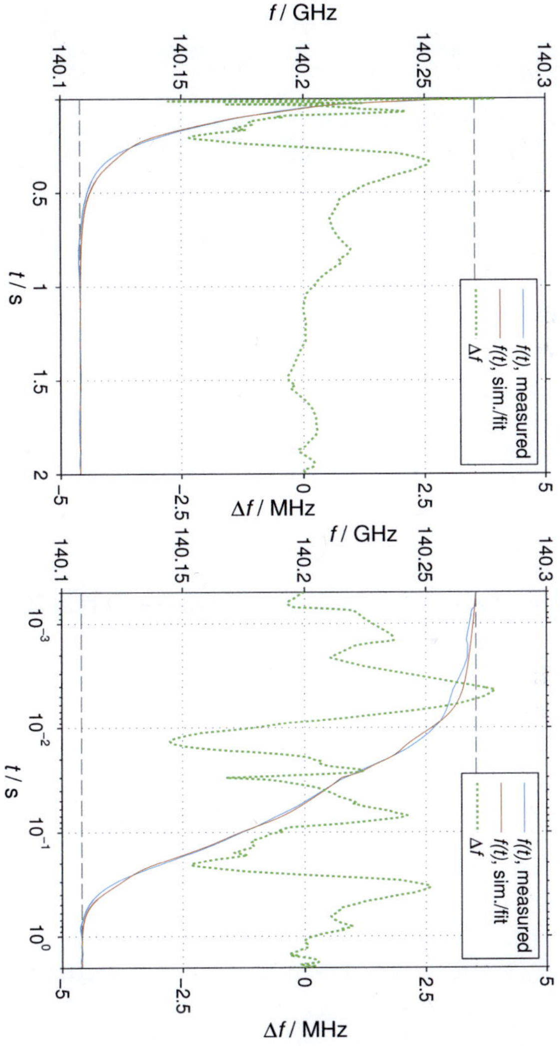

Figure 7.8: Least-squares smoothed measured frequency with global fit result as marked in Fig. 7.7, and difference between the curves in linear and logarithmic time scale

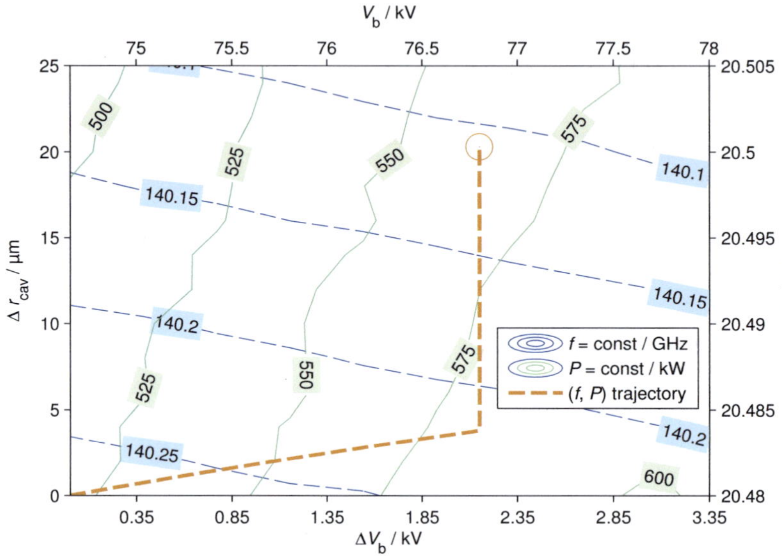

Figure 7.9: Parameter-map with final trajectory for a 500 kW pulse

Discussion

With $|\Delta f| < 5\,\mathrm{MHz}$, the fit quality obtained in the global maximum is at the limit of the quality of the input data, both from the numerical as well as from the experimental side:

- In Section 3.7, typical deviations in simulation up to 20 MHz were obtained. Over ranges of a few kV, the deviations from measurement remained quite constant, so the consequence would be a mostly static offset, which is taken into account in the fitting process.
- While with the PSA system the experimental data can be obtained down to kHz-range resolutions, the gyrotron frequency itself is unfortunately less well defined. As demonstrated in Sections 5.1.2 and 5.1.3, frequency jumps up to 10 MHz occur.

Hence, it can be stated that the best reasonably possible fit result has been obtained, and that the data have no room for further optimization. Especially there is no optimization potential left to test more complex alternative time dependence formulations.
The comparison of the fit results with the time constants obtained from calculation raises questions, since the fitted neutralization time constant of 30 ms disagrees significantly with the calculated value of 400 ms from (7.16). Interestingly, the theoretically obtained value corresponds reasonably well to the secondary fit minimum at 590 ms. This can be interpreted as an indicator for the presence of an additional time dependence.[26] A probable explanation can be found in the outgassing, which was neglected in the base function definition.

7.3.4 Outgassing revisited

Even in dedicated vacuum vessel experiments with tight temperature control, the outgassing remains hard to characterize. Thus, as stated before, the time dependence and absolute value of the term $q_{\mathrm{N}}^{\mathrm{w}}$ in (7.14) are unknown. In the following, therefore, only a most simple first-order modification of the description is used, in order to derive a general notion of the outgassing consequences.
As the fitting technique residuum has no improvement reserve, the results obtained cannot be verified this way.

Crude modeling

In a well conditioned gyrotron, local outgassing in the first few seconds will be dominated by physisorbed particles. Because of the skin effect, all RF losses are deposited in the wall surface, to which a physisorbed layer is necessarily in direct contact. A comparison of the possible desorption energies (compare [Jou08]) with the magnitude of the known RF loss power indicates that it will evaporate very quickly. Chemisorbed and absorbed particles can be expected to take longer

[26]Note that also in [DBI+96] two different time scales were used for the neutralization process, without further discussion.

to be desorbed, and hence are less relevant at the very start of the pulse. For the sake of simplification, the surface layer can be assumed to desorb instantaneously at $t_\mathrm{p} = 0^+$. This allows one to drop the term q_N^w from (7.14), and turn the local outgassing contribution into an instantaneous density increase at the start of the pulse.[27]
To quantify this instantaneous gas desorption in a very crude estimation, the measured pressure rise in the mirrorbox-collector range during the first few seconds of a pulse could be used: Assuming that it comes only from the described physisorbed particles, desorbing more slowly than in the cavity since the average power densities on the wall are much smaller in the mirrorbox-collector vessel, a maximum "outgassing factor" of $a_\mathrm{vess} = 2$–5 for the vessel density increase at the start can be assumed (compare the sample curves in Fig. 7.2).
Both the CPW area and the mirrorbox-collector vessel can be approximated as cylinders, with a uniform physisorbed surface particle layer before the start of the pulse. Particle density is a volume-based quantity, while the amount of physisorbed gas depends on the surface area. Hence a_vess must be scaled with the radius ratio of the two vessels, to express the density increase in the cavity potential well area:

$$a_\mathrm{cav} \approx \frac{r_\mathrm{vess}}{r_\mathrm{c}} a_\mathrm{vess} \approx 10\, a_\mathrm{vess} \approx 20 - 50 \quad . \tag{7.22}$$

This means that at pulse start, the neutral particle density in the cavity is expected to increase by roughly one order of magnitude.
The net flow $q_\mathrm{N}^\mathrm{vess}$ is actually a superposition of two flows with opposite directions

$$q_\mathrm{N}^\mathrm{vess} = (n_\mathrm{vess} - n_\mathrm{cav}) C_\mathrm{t} \quad , \tag{7.23}$$

governed only by the particle density difference between the two compartments and the properties of the connector.
From the vessel size ratio consideration above, it can be deduced that the "fast outgassing" increases n_cav considerably more than n_vess, because $a_\mathrm{vess} < a_\mathrm{cav}$. At the start of the pulse, this would cause an

[27]Note that this is in conflict with the conductivity-based static pressure differences used in Section 7.2.5, as it removes the system from equilibrium.

initial net flow from the cavity to the mirrorbox, which superposes with the ionization process. After a short amount of time during which $n_{\mathrm{g}}(t)$ has an exponential time dependence (compare Section 7.2.6), the locally outgassed particles neutral have been dissipated, and the linear process as expressed in (7.21) dominates.
Inserting (7.23) into (7.14) yields a modified flow equation:

$$\frac{\partial N_{\mathrm{cav}}}{\partial t} = n_{\mathrm{vess}} C_{\mathrm{t}} - n_{\mathrm{cav}} C_{\mathrm{t}} - g_{\mathrm{i}} N_{\mathrm{cav}}(t) \tag{7.24}$$

$$= n_{\mathrm{vess}} C_{\mathrm{t}} - \left(\frac{C_{\mathrm{t}}}{V_{\mathrm{cpw}}} + g_{\mathrm{i}} \right) N_{\mathrm{cav}}(t) \quad . \tag{7.25}$$

Further simplifying, it is also assumed here that $n_{\mathrm{vess}} = \mathrm{const}$, which is justified by the outgassing factor relations (7.22). The solution to this first order inhomogeneous differential equation with the starting value $N_{\mathrm{cav}}(0) = a_{\mathrm{cav}} n_{\mathrm{cav},0}$ is

$$N_{\mathrm{cav}}(t) = \frac{n_{\mathrm{vess}}}{A} + \left(a_{\mathrm{cav}} n_{\mathrm{cav},0} V_{\mathrm{cpw}} - \frac{n_{\mathrm{vess}}}{A} \right) e^{-At} \quad , \tag{7.26}$$

using the abbreviation

$$A = \frac{C_{\mathrm{t}}}{V_{\mathrm{cpw}}} + g_{\mathrm{i}} \quad .$$

From the now available explicit formulation of the gas particle time dependence $N_{\mathrm{cav}}(t)$, the ion particle development can be derived by integration of

$$\frac{\partial N_{\mathrm{i}}}{\partial t} = g_{\mathrm{i}} N_{\mathrm{cav}}(t) \quad . \tag{7.27}$$

With the second abbreviation

$$B = n_{\mathrm{vess}} C_{\mathrm{t}}$$

the neutralization function $\zeta(t)$ for $\zeta < \zeta_{\mathrm{end}}$ can with (7.8) be directly expressed as

$$\begin{aligned} \zeta(t) = \frac{g_{\mathrm{i}}}{A^2 N_{\mathrm{b}}} \{ & (B - A a_{\mathrm{cav}} N_{\mathrm{cav},0}) \, e^{-At} + \\ & ABt + A a_{\mathrm{cav}} N_{\mathrm{cav},0} - B \} \quad . \end{aligned} \tag{7.28}$$

To illustrate the effect of this modification on the time-dependent behavior, Figure 7.10 contains sample $\zeta(t)$ curves for different values of a_{cav}. For $a_{\text{cav}} = 1$, the purely linear development as used in the studies above is observed[28], while higher outgassing factors lead to increasingly pronounced initial exponential behavior. Consequently, $a_{\text{cav}} > 1$ means faster neutralization in the early part of the pulse, and a corresponding downward shift of the total neutralization time t_{n}, as indicated by the sample interception markers at $\zeta = \zeta_{\text{end}} := 0.6$.

7.3.5 Sample results

The effect of the new description can again be assessed in the RMS map, an updated version of which with $a_{\text{cav}} = 10$ is given in Figure 7.11. It shows a clear unification of the previously disparate minima into a single global maximum at $11\,/\,\text{s} < \lambda_{\text{th}} < 13\,/\,\text{s}$ and $25\,/\,\text{s} < \lambda_{\text{n}} < 40\,/\,\text{s}$. It should be noted again that this is a qualitative result, since the total residuum ϵ cannot be significantly improved over the previously obtained results. Thus, a further optimization with the additional free parameter a_{cav} was also not conducted, which is why the optimum obtained here is slightly less good than the one found in the previous simple formulation.

Note that the interpretation of $1/\lambda_{\text{n}}$ as a time constant is still valid, but one must always consider the underlying function (7.28), which is now more complex. The process is no longer linear, so the neutralization time t_{n} differs from the time constant τ_{n} and must be calculated by solving $\zeta(t) = \zeta_{\text{end}}$. For example, with $a_{\text{cav}} = 10$ and $p = 1.8 \cdot 10^{-9}\,\text{mbar}$, the value $t_{\text{n}} = 370\,\text{ms}$ is obtained.

Discussing Figure 7.8 in Section 7.3.3, it was noted that the neutralization time obtained with the linear fit corresponded well with the kink in the measured frequency at $t = 30\,\text{ms}$. The formulation with outgassing has a considerably longer neutralization time, but interestingly, as visible in Figure 7.10, the $\zeta(t)$ here has the crossover point from exponential to linear development, and thus its own rather sharp change in time dependence.

[28] Note that this result also justifies the selection of case 1. in 7.2.6.

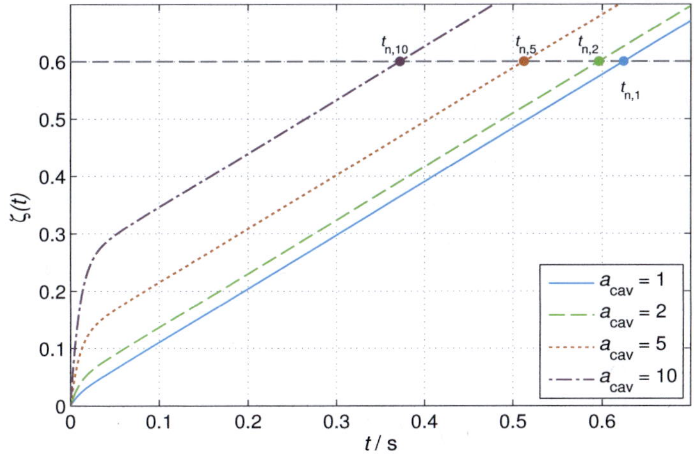

Figure 7.10: $\zeta(t)$ for different a_{cav} values at $p_{\text{cav},0} = 1.8 \cdot 10^{-9}$ mbar

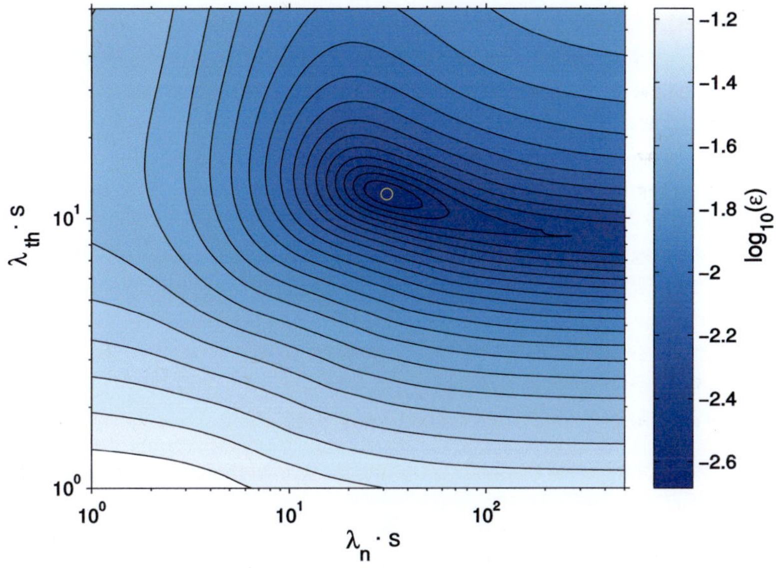

Figure 7.11: Sample RMS map for outgassing-corrected $\zeta(t)$ formalism

7.4 Summary

A simplified formulation to describe the frequency tuning caused by electrostatic neutralization was developed. For this, both the ionization-related quantities as well as the neutral gas conditions inside the gyrotron were discussed. While the derivation of the ionization rules could be carried out in a rigorous and straightforward manner, the neutral gas dependences introduce the highest uncertainty, mainly due to hardly controllable outgassing. However, it should be pointed out that it is indeed the consideration of the neutral gas phenomena which had considerable impact on the results.
As the neutralization process itself and the directly related quantities cannot be monitored externally, a fitting technique which makes use of the neutralization influence on the RF frequency was developed and applied. This way, a verification of the theoretical considerations based on the quantity which can be externally measured with the highest accuracy, the frequency, was made possible.
Already using first order formalisms, a significance limit for the fitting technique was reached. An extended formulation, taking into account not only the gas dynamics but also basic outgassing considerations, could thus be derived but not finally verified.
The most important findings of the theoretical and the numerical investigations can be listed as follows:[29]

- In agreement with electrostatic confinement considerations, the results indicate that even in LP steady state, complete neutralization is not achieved, and V_b will not be equal to the external acceleration voltage at any point of time. The remaining space charge depression voltage in the case of the W7-X gyrotrons can be estimated to be approximately $0.4V_\mathrm{dep}$.

[29] The investigations were made for 500 kW pulses, as in this operation very suitable measurements could be made. The conclusions also apply to operation at 1 MW. Note, however, that in Section 7.2.6 the case selection becomes problematic then, as with increased beam current the ionization rate increases, causing $q_\mathrm{N}^\mathrm{i} \approx 1.5 q_\mathrm{N}^\mathrm{vess}$. This can cause a slight difference, but only if absolutely no outgassing is assumed to take place.

- For technically achievable very low tube pressures, the neutralization process is not limited by the amount of available neutral gas in the tube, and will not influence the total pressure measurably. Especially no "pump down" by the electron beam, acting as a sink for a significant amount of neutral particles, is possible. The main reason for this is the large total volume of the vessel, which allows internal gas flows to replenish the ionized particles through constant gas-dynamic flows. An additional important factor are the wall outgassing processes.
- Following the gas dynamics considerations, the neutralization process is likely a mostly linear function of time, which in first order can be approximated by (7.21) with $\lambda_{\mathrm{n}} = 1/g_{\mathrm{i}}$ from (7.11). This is only an upper limit for the neutralization time, as a higher process rate at pulse start due to local outgassing, having exponential characteristics, can be expected. As a consequence, simple fitting approaches can encounter multiple fit optima, or yield ambiguous results depending on the chosen time dependence.

This investigation showed that the frequency tuning process at the start of long pulses depends on multiple properties of the gyrotron design, for example the shape of the RF geometry, and a number of parameters which are not directly determinable, e.g. the outgassing behavior. This prevents the formulation of a general analytical model for simple application to other gyrotrons.

8 Conclusions

8.1 Summary

A wide range of transient effects in gyrotrons was investigated, with a focus on gathering the primary information from the RF output spectrum. For this, a high-resolution high-dynamic range pulse spectrum measurement system (PSA, see Chapter 4) was developed, which combines time-dependent analysis and frequency unambiguity in the entire D-Band. With 60 dB of dynamic range with an instantaneous bandwidth of 6 GHz per receiver channel pair, this establishes a new state-of-the-art in the field of gyrotron-related spectral measurement. The system was used to gather spectral information of unprecedented quality and detail, enabling new paths to the investigations of the inner physics of the gyrotron especially by its ability to retain single-shot RF frequency unambiguity. In Chapter 5, the unique capabilities of the system were exemplified in a number of measurement examples, covering four different gyrotron types in the range 118–170 GHz and a wide variety of effects, which were discussed in the context of modern gyrotron theory.

The connection between experimental and simulation data was made with a rigorous and detailed study, in which first of all the comparison framework for simulated and measured RF powers in short pulse operation of W7-X gyrotrons was established in Chapter 3. It was demonstrated that the first few milliseconds of a pulse constitute a quasi-stationary operation, which is suitable for systematic and efficient gyrotron investigation. Several versions of the leading European gyrotron interaction simulation codes SELFT and EVRIDIKI were compared systematically in a benchmark with unprecedented breadth, to the best knowledge of the author. While over a wide parameter range, very good agreement concerning frequency and good agreement in power within the expected uncertainty was achieved, the reproduction of the cavity mode transition was found to be problematic for all tested codes. The entire undertaking also emphasizes the importance of using wide parameter sweeps instead of single or sparse operating points, as typically practiced in the gyrotron community. Furthermore, detailed consideration was given to analytical linear-theory approaches, for which no published quantitative assessments seem to be available.

An extensive study of parasitic RF oscillations in W7-X gyrotrons was conducted, in the course of which a general classification scheme for these effects was suggested. A number of paths for systematic theoretical and experimental analysis were developed and applied. The experimental investigations included wide parameter sweeps, in combination with cavity mode transition and parasite starting beam current investigations, all of which were carried with the help of the new PSA system. Only the fusion of the data gathered with the various approaches, including the meticulous application of large-scale interaction simulations to a large number of experimentally characterized operating points, lead to the diagnosis of after-cavity interactions (ACI) in the launcher area. The existing limitations of slow-time and quasi-PIC gyrotron interaction simulation methods and the related modeling for the investigation of ACI were explored and discussed; especially the failure to reproduce parasite starting beam currents shows that the existing interaction codes are not suited to treat all aspects of the ACI reliably. Whether this is only due to the known physical limitations of the interaction simulation codes or also a problem of the used beam parameters, which were simulated separately, is to be resolved. The associated parasitic modes in the presented study could not be identified with absolute certainty, however, the most probable ones responsible for the observed undesired signals in the general frequency range 130.2–132.7 GHz are the modes $TE^{-}_{28,8}$ and $TE^{+}_{25,9}$, the former being the nominal cavity mode. To date, the presented work might be the strongest indication for the physical existence of ACI. The study also underlines the necessity for solid and extensive experimental investigation, as from this source the most significant indicators for the decision between BCI and ACI were gathered.

In order to establish a more clear picture of the transient state in which the gyrotron is between pulse start and the first few seconds of operation, the frequency tuning processes at the onset of long pulses were investigated. The compilation of the determining factors of the neutralization process showed that detailed and realistic information on the ion confinement properties and the kinetic gas dynamics inside the tube is necessary for a proper description of the situation. It was found that while the thermal expansion process is of exponential character, the neutralization process is primarily linear. A macroscopic formulation of the probable time dependences, neutralization rates and final neutralized depression voltage fractions was introduced for the experimental situation of W7-X gyrotrons. This permitted a fitting technique by which the theoretical predictions could be connected to experimental information about the transient status of the gyrotron. The results indicate that in LP operation, the voltage depression is never fully neutralized, but only up to a fraction of $\sim 60\%$. Further, in contrast to long-time assumptions, the neutralization time in modern high power gyrotron was found to be of similar order or shorter than the thermal cavity expansion time, and to very probably follow a first exponential, then linear dependence influenced by local outgassing and internal gas flows.

Summarizing, a number of relations between the domains of simulation and experiment were verified or newly established. This yields new ways of investigating and explaining the behavior of the gyrotron, especially concerning transient states of operation, for which the diagnostic possibilities were drastically extended.

8.2 Outlook

All topics contained in the work have potential for further development and extension. In the following, only the most obvious next steps for the PSA system, the problem of parasitic oscillations and the investigations of long pulse effects are listed.

PSA system

In its current version, the PSA system provides 7, resp. 14 ms (using 2, resp. 4 hardware channels) of unambiguous measurement with a total span of 6 GHz per channel pair, and undersampled continuous acquisition up to 2 s. A logical further improvement would be the implementation of the spectrum analyzer emulation technique mentioned in Section 4.7.2, as it would further increase the system's flexibility. Periodic coverage with intervals of e.g. 0.5–2 s would then allow stationary long pulse investigations with frequency resolutions similar to those in the regular spectrograms, maintaining RF unambiguity.

The reconstruction technique as it is implemented only validates the pre-defined "desired acting harmonics", while discarding all information received through other harmonics. Of course for each configuration, there are also other Δf values which correspond unambiguously to other harmonic pairs. This could be used to reconstruct further RF spectrograms from the saved data, extending the system's frequency coverage. While these additional spectrograms would have a lower individual bandwidth than the primary ones, the total frequency coverage would be extended without further hardware investment.

A PSA potential which has remained mostly untapped is the possibility of phase analysis on the basis of the acquired signals. The extraction of modulation waveforms or even full demodulation could give novel insights concerning the field of modulating LF oscillations. With help of the high sampling rate employed, even full IQ (inphase-quadrature) demodulation could be software-implemented: Following a suggestion in [Con83], this can be done using only one channel without any change in the setup, however, sacrificing 75% bandwidth.

For routine use, a number of technical improvements can also be addressed. The wealth of data created must be turned into information, which is an often underestimated problem. In particular the compilation of a survey of measured frequencies over a long period of time, or the question under which circumstances a specific frequency has been experimentally observed, are problems which can consume high amounts of operator time. A solution could be found in an extension

of the frequency tracking technique introduced in Chapter 4: It could form the basis of a fully-automatic program such as the "thumbnail-generators" of modern image management programs; it could browse and index all stored spectrogram data once, extracting and collecting frequency information for a database. This would facilitate scatter-analyses such as the one presented in Figure 6.6 tremendously, and give way to high level comparisons between different gyrotrons.
Also, the integration of the PSA system into the gyrotron test stand is a multi-faceted problem, as operational flexibility of the other measurement systems must be maintained.

Simulations

In this work, only the simplified RMS beam parameter spread definition was employed, while as noted in Section 2.1.1, statistically more flexible descriptions like the relative spread formalism exist. Also it is possible to export the beam parameter distributions from the electron-optical simulation tools to be used directly in the interaction simulations, removing the shortcomings of statistical descriptions altogether. An according expansion of the benchmark simulations of Chapter 3 could be highly instructive, and also have impact on the simulation of parasitic oscillations as conducted in Section 6.4.
The comparison of linear frequency calculations with the results of simulated and measured values showed significant differences. It would be very useful to obtain an analytic and precise frequency estimation method based on the geometry and the beam parameters, to enhance techniques such as the Brillouin analysis in Chapter 6, and greatly simplify investigations as the one in Chapter 7.

Parasitic oscillations

Various approaches for parasite investigation have been discussed and applied, but the problem remains complex and elusive. New and preferably independent analysis tools would be highly welcome, so new paths along which these might be developed have to be found.

Further steps in this area also include investigating means to suppress the characterized parasites, which of course would primarily involve theoretical considerations and simulations. The need for a non-experimental validation of suggested improvements, however, leads directly to the general design problem of parasite prediction, for which no reliable and general solution currently exists.
A large potential lies in full-PIC simulations, as they are capable of not only inherently considering all modes, but also to correctly model the launcher with all its asymmetrical properties. To revisit the simulations made in Section 6.4 with such a tool could be the only way of checking the conclusions reached in Chapter 6, and possibly also to solve the design problem of parasite prediction mentioned above.

Thermal expansion and neutralization

The results could be used in parameter studies, and be compared with thermo-mechanical simulations of the cavity expansion. Uncertain material parameters could be varied in these simulations to gain result matches of the resonator expansion times obtained with the fitting method, narrowing down the probable ranges of these parameters.
Of course the remaining, and considerable, uncertainties must be kept in mind. From the means developed and applied in this work, more improvement cannot be expected, as the achievable result certainty has already been reached.
An important point which could add to the knowledge about the processes is the fate of the background electrons, i.e. the low-energy electrons which are released in an ionization process from neutral particles. Being trapped because of their low kinetic energy they might also modify the electrostatic profile and contribute to further ionization, dynamically altering the conditions. This could be done by using more sophisticated means such as ARIADNE simulations. Also the exact ion locations and their axial motion were not considered, and could be investigated in the same way: Depending on the location of their origin, they oscillate in the CPW area, and therefore contribute only partially to the actual neutralization potential inside the RF cavity.

A Digital signal analysis by STFT

The application of the Short-Time Fourier Transform (STFT) to sampled signals is a key feature of the PSA system presented in Chapter 4. In the following, the details of this operation are explained with their relevance to the measurement system.
More elaborate accounts on this can be found in [Pro07], and more application-oriented knowledge in [Lyo11].

A.1 Fourier transform of discretized signals

The general Fourier transform

$$s(t) \quad \circ\!\!-\!\!\bullet \quad \mathcal{F}\{s(t)\} = \int_{-\infty}^{+\infty} s(t)\mathrm{e}^{-\mathrm{j}2\pi f t}\,\mathrm{d}t \tag{A.1}$$

is applied to a continuous signal of unlimited length $s(t)$; this yields infinite frequency resolution.
Real-world signals in digital systems are discretized in value and time, and never of infinite duration. This corresponds to a periodic sampling of $s(t)$ with the constant sampling interval t_r, and can expressed mathematically as the multiplication with a t_r-periodic Dirac pulse train

$$\sum_{n=-\infty}^{+\infty} \delta(t - nt_\mathrm{r}) \quad . \tag{A.2}$$

The finite acquisition time t_a corresponds to the selection of a signal interval through multiplication with a rectangular function

$$rect_{t_\mathrm{a}}(t) = \begin{cases} 1 & \text{if } 0 \le t < t_\mathrm{a} \\ 0 & \text{overwise} \end{cases} \quad . \tag{A.3}$$

This leads to a finite number of samples $N = t_\mathrm{a}/t_\mathrm{r}$:

$$s_\sqcap(n) = s(t) \cdot rect_{t_\mathrm{a}}(t) \cdot \sum_{n=-\infty}^{+\infty} \delta(t - nt_\mathrm{r})$$

$$= s(t) \sum_{n=0}^{N-1} \delta(t - nt_\mathrm{r}) \quad . \tag{A.4}$$

Applying (A.1) to this time-discrete signal yields

$$S_\sqcap(t) = \mathcal{F}\{s_\sqcap(n)\} = \int_{-\infty}^{+\infty} s(t) \sum_{n=0}^{N-1} \delta(t - nt_\mathrm{r}) e^{-\mathrm{j}2\pi f t}\,\mathrm{d}t$$

$$= \sum_{n=0}^{N-1} \int_{-\infty}^{+\infty} s(t)\delta(t - nt_\mathrm{r}) e^{-\mathrm{j}2\pi f t}\,\mathrm{d}t$$

$$= \sum_{n=0}^{N-1} s(nt_\mathrm{r}) e^{-\mathrm{j}2\pi f n t_\mathrm{r}}. \tag{A.5}$$

This term describes a continuous spectrum of infinite bandwidth, in correct mathematical accordance with the discrete and time-finite nature of the input signal.
However, the fundamental reciprocity of time and frequency yields the expression of the smallest describable frequency as $f_\mathrm{a} = 1/t_\mathrm{a}$, which defines a sensible frequency discretization

$$f_k = k \cdot 1/t_\mathrm{a} \quad , k \in \mathbb{Z}. \tag{A.6}$$

The insertion of this definition together with $t_\mathrm{a} = N \cdot t_\mathrm{r}$ into (A.5) yields the discrete Fourier transform (DFT)

$$\mathcal{F}\{s_\sqcap(n)\} = \sum_{n=0}^{N-1} s(nt_\mathrm{r}) e^{-\mathrm{j}2\pi k \frac{n}{N}} \tag{A.7}$$

with a purely numerical and discrete angular frequency $2\pi k$.

A.1.1 Aliasing

By definition there is no maximum value for k in an infinite spectrum; a practical limit is found over the N-periodicity of the complex exponential, as for example, the angular frequencies in $-N/2 < k \leq N/2$ are identical to the ones in the range $N/2 < k \leq 3N/2$.
This can also be observed more mathematically already from the sampling action in the time domain: The Fourier transform of (A.4) denotes a convolution operation with the transform of the Dirac pulse train, indicating a periodic reproduction of $\mathcal{F}\{s(t)\}$ with $1/t_\mathrm{r}$. If the spectrum $\mathcal{F}\{s(t)\}$ has a higher bandwidth than $1/(2t_\mathrm{r})$, the spectral aliases will overlap. The spectrum is therefore strictly periodic, which yields a practical span of N adjacent values for an arbitrary k range. In the spectrum, we thus find the same number of relevant bins[1] as the input values that were used.
This periodicity introduces the Nyquist frequency $f_\mathrm{N} = 1/(2t_\mathrm{r})$, or numerically $k_\mathrm{N} = N/2$, whose integer multiples are used to divide the frequency range into so-called *Nyquist zones*, compare Figure A.1.
From the previous discussion, it becomes clear that an incoming signal with an arbitrary frequency $f > f_\mathrm{N}$ which reaches the A/D-converter will appear at the minimum alias frequency

$$f_\mathrm{d} = \left| 2\lfloor \frac{f + f_\mathrm{N}}{2f_\mathrm{N}} \rfloor f_\mathrm{N} - f \right| \quad , \tag{A.8}$$

that means mirrored by f_N in the first Nyquist zone.[2]

[1] A "bin" is one value element along a discrete time or frequency axis.
[2] The expression $\lfloor \cdot \rfloor$ denotes the *floor()*-function, i.e. next lower integer of the argument.

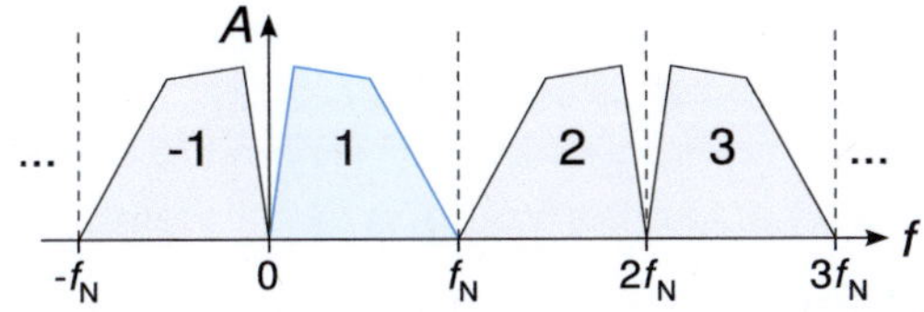

Figure A.1: Illustration of different aliasing/Nyquist zones

In technical applications, typically a low-pass filter is employed at the A/D-converter input to limit the effective acquisition to the unambiguous range $f < 1/2t_{\mathrm{r}}$, i.e. the first Nyquist zone.

A.1.2 Power spectra

In (A.5), another important property of the Fourier transform is visible, which is also illustrated in Figure A.1:

$$S_{\sqcap}^{*}(f) \equiv S_{\sqcap}(-f) \;\forall\; s(t) \;\in\; \mathbb{R}. \tag{A.9}$$

This is a symmetry relation, which indicates that for a real-valued time signal the complete attainable information resides duplicated in each half of a spectrum with the bandwidth $2f_{\mathrm{N}} = 1/t_{\mathrm{r}}$.
To analyze the frequency content of a signal, usually the power density spectrum is implemented as

$$S(f) = \frac{|2\mathcal{F}\{s(t)\}|^2}{Z_0} \quad \text{with } f \geq 0 \quad . \tag{A.10}$$

This can be directly related to a physical power which is absorbed by reference impedance Z_0 in a norm bandwidth. In the following and the rest of this work, unless otherwise indicated, the term *spectrum* refers to this quantity, using $Z_0 = 50\,\Omega$.

A.1.3 Window functions

As noted before, measured signals have always finite length and sampling rate, as was expressed through the multiplication of $s(t)$ with $rect_{t_{\mathrm{a}}}(t)$ in (A.4). Mathematically, the multiplication of time signals is equivalent to the convolution of the respective Fourier transforms. $\mathcal{F}\{rect_{t_{\mathrm{a}}}(t)\}$ is the cardinal sine, $\sin(f)/f$.
As long as the frequency content in $s_{\sqcap}(t)$ is equal to a discrete frequency k in $S_{\sqcap}(f)$, the effect of this convolution is not visible in the spectrum. However, for input signals with $f \neq k/t_{\mathrm{a}}$, the spectral is energy spread over multiple frequency bins, leading to unwanted distortions in level display and frequency content. This is called *leakage*.

In order to counteract the problematic rectangular filtering, a scaling function $w(t)$ can be applied to the input time data, generally referred to as a Fourier window. Window functions are normally symmetric, real-valued and have the same numerical length N as the input time signal.
From the large number of window functions which were characterized in the past [Har78], a suitable choice can only be made with an eye on the intended application.
Typical optimization topics are:

- Maximum scalloping loss, which is the relative loss in spectral amplitude through leakage.
- Side lobe levels and their roll-off; side lobe control can be crucial to prevent masking of adjacent weak signals through the side lobes of a strong signal.
- Main lobe width, which is larger than the one of the *sinc()* function for all other windows and degrades the frequency resolution.

The main lobe width of a window invariably deteriorates the frequency resolution, and causes the analysis to have an effective resolution $\delta f_{\mathrm{eff}} = r \cdot \delta f_{\mathrm{num}}, \quad r \geq 1$. Typical values for r are in the range 1.1–2. For broadband investigations this is of little impact, which is why in the following and in the rest of this work only df_{num} is used for simplicity.

A.2 Dynamic range

A.2.1 Time domain

The basic framework of the achievable dynamic range is already set at the time-domain signal acquisition:[3] Typically, a sample-and-hold circuit with a vertical resolution of b bits delivers signal values with the sampling rate $f_{\mathrm{r}} = 1/t_{\mathrm{r}}$.
For an ideal A/D-converter with a perfectly stable sampling rate, the dominating quantity for the achievable dynamic range is b. The total

[3] In the following, we use and generalize the derivation in [Lyo11].

input range $-V_0 \dots V_0$ is then quantized with 2^b values, which yields a minimum value difference, or discretization step height, of

$$q = \frac{2V_0}{2^b} = \frac{V_0}{2^{b-1}} \quad . \tag{A.11}$$

In order to take into account the saturation of the A/D front-end by the input signal amplitude V_s, a load factor $L = V_\mathrm{s}/V_0$ can be defined. The overall achievable SNR is then expressed as the ratio of the input power to the power of the quantization-induced noise floor:

$$SNR_\mathcal{T} = \frac{P_\mathrm{s}}{P_\epsilon} \quad . \tag{A.12}$$

Assuming an even quantization error distribution yields a uniform probability density function $f(\epsilon)$ with the value $f = 1/q$ inside the limits $|\epsilon| \leq \frac{q}{2}$. The average background noise power is therefore[4]

$$P_\epsilon = \sigma_\epsilon^2 = \int_{-q/2}^{+q/2} \epsilon^2 f(\epsilon) \mathrm{d}\epsilon = \frac{q}{12} \quad . \tag{A.13}$$

With the power $P_\mathrm{s} = V_\mathrm{s}^2/(2Z)$ of a sinusoidal input signal and the definition of the load factor, a linear SNR of

$$SNR_\mathcal{T} = \frac{P_\mathrm{s}}{P_\epsilon} = \frac{3}{2} \cdot \frac{L^2}{Z} \cdot 2^{2b} \tag{A.14}$$

can be achieved. Expressed in decibels, the formula becomes

$$SNR_\mathcal{T}/\mathrm{dB} = 10 \log_{10} \frac{3}{2} + b \cdot 20 \log_{10} 2 - 10 \log_{10} Z + 20 \log_{10} L \quad . \tag{A.15}$$

Although clearly a load factor $L = 1$ would be optimal, it is a case that is technically not desirable: For $L > 1$, signal clipping would

[4]This does not have to be a physical noise floor, i.e. similar in appearance to white Gaussian noise. Depending on the input signal, correlated noise can create a discrete and highly inhomogeneous spurious spectrum. The usable dynamic range is therefore often not defined by a continuous noise floor, but by the level of the strongest spurious line, leading to the usage of the property SFDR.

occur, which would render the acquired signal useless by introducing irrecoverable distortions. Therefore typically load factors -6 dB to -3 dB are employed.[5]
In connection with time discretization, a further SNR limitation arises: The sampling rate of a real A/D converter has a limited short-time stability, which can be expressed as phase noise or jitter. This phenomenon also creates numerical noise, since it causes the samples processed by the DFT to be not perfectly equidistant [Pro07]. Typically it is expressed as a degradation of the usable A/D converter vertical resolution, in the form of the *effective number of bits*, ENOB.
With the help of certain assumptions on the jitter statistics, ENOB can be calculated analytically [Kes08] and is found to be connected reciprocally to the usable signal bandwidth. Manufacturers normally measure it directly during device characterization and give the value for the specified bandwidth. This technique also allows the consideration of additional effects such as A/D converter nonlinearities [Sch11].

A.2.2 Frequency domain

The transition to the frequency domain via the Fourier transform yields the benefit of so-called *processing gain*, which is generally expressed as an addition to (A.15) as

$$SNR_{\mathcal{F}} = SNR_{\mathcal{T}} + 10\log_{10}\frac{N}{2} \quad . \tag{A.16}$$

This effect is often explained in the frequency domain, by analogy of the DFT to a filterbank: Reducing the channel width by a factor of two will gain 3 dB of dynamic range if the background noise is Gaussian. In the time domain, the correspondence is found in the implicit averaging process of the DFT. This is visible in (A.7), as the value for each frequency k is the result of an implicit averaging over all N input values. As long as the input samples are statistically independent, the amplitude uncertainty lowers with $1/\sqrt{N}$.

[5] Equation (A.15) reduces to the standard form $SNR/\mathrm{dB} = 6.02 \cdot b + 1.76$ propagated in most text books only for $Z = 1\,\Omega$ and $L = 1$.

Together with the effective resolution of the A/D converter, the DFT processing gain is the dominant factor in many technical signal analysis situations. As an example, for a time signal acquired with 10 GS/s, a 1 MHz resolved DFT requires $N = 10^4$, yielding a processing gain of 37 dB.

A.3 Time-frequency analysis

A.3.1 General considerations

Popular techniques for time-frequency signal analysis are the *short time Fourier transform* (STFT), the Wavelet transform and the Wigner-Ville distribution. Wavelet analysis includes the application of a time shifted and scaled kernel function, the wavelet. It therefore inherently includes multi-resolution analysis which has wide applications in image processing, pattern recognition and compression, but would induce limited comparability with other techniques when applied to microwave signals. The STFT can be interpreted as a special case of the wavelet transform with the restriction of harmonic basis functions and fixed scale.

The Wigner-Ville distribution is an approach closer to more abstract signal processing problems, and allows more degrees of freedom than the STFT. However, it is subject to inherent cross-correlation ambiguity terms, which could complicate or compromise the ambiguity goals formulated in Section 1.4 of this work. More information on the different approaches can be found for example in [Coh89].

For this work, the STFT was chosen for signal analysis, and will in the following be developed from the previously introduced DFT.[6]

[6]Instead, the STFT can also be derived directly from the continuous domain, with time and frequency being exchangeable. This more general approach, the Gabor transform, yields the same result.

A.3.2 Discrete short-time Fourier transform

The application of the DFT as introduced in Section A.1 on M segments of the acquired time-discrete signal with $t_{\mathrm{r}} \ll t_{\mathrm{seg}} \ll t_{\mathrm{a}}$ will yield M discrete real-valued power spectra S_i. The segment selection and the Fourier windowing in the time domain can be expressed in one step through the repeated application of a time-shifted Fourier window to the original signal as

$$s_w(t, \tau_i) = s(t) \cdot w(\tau_i - t) \quad , i = [0, M]. \tag{A.17}$$

Each segment is now characterized by a segment time index τ_i and the ordinary time coordinate $0 \leq t < t_{\mathrm{seg}}$.
Applying the Fourier transform and the power spectrum definitions from Section A.1 to each time signal $s_w(t, \tau_i)$ yields i individual spectra $S(f_k, \tau_i)$, each with a numerical frequency resolution $\delta f = 1/t_{\mathrm{seg}}$ and a bandwidth of $\Delta f_{\mathrm{IF}} = f_{\mathrm{r}}/2$. By concatenation of these spectra along the τ axis, a two-dimensional matrix representing the spectral evolution over time is obtained, with the indices k and i for frequency and time. This is typically referred to as a *spectrogram*, *waterfall diagram* or simply STFT plot and represents a time-dependent power spectrum.
As the arbitrary but fixed repetitive shift by $\delta\tau = \tau_i - \tau_{i-1}$ of the operation leads to a discrete numerical time step between the resulting spectra, it can be treated like a true time axis, as long as the background of its creation is kept in mind. For simplicity, the time axis in spectrograms will hence be denoted by the ordinary time variable t during the rest of this work.

A.3.3 STFT time step / overlap factor

The number M of individual spectra depends on both δt and t_{seg} in relation to the total available length of the input signal t_{a}. It is practical to link these parameters through an overlap factor

$$x_{\mathrm{OV}} = 1 - \frac{\delta t}{t_{\mathrm{seg}}} \quad . \tag{A.18}$$

Applying the basic restrictions $0 < \delta t \leq t_{\mathrm{seg}}$ gives an overlap factor range of $x_{\mathrm{OV}} \in [0, 1)$, with $x_{\mathrm{OV}} = 0$ denoting non-overlapping adjacent segments with $\delta t = t_{\mathrm{seg}}$ and $M = \lfloor t_{\mathrm{a}}/t_{\mathrm{seg}} \rfloor$.
The degradation of the frequency resolution through the Fourier window also has a similar but reciprocal effect in the time dimension. So in addition to the previously defined effective frequency resolution δf_{eff}, an effective spectrogram time resolution δt_{eff} is defined. The former is transported by the main lobe width of the window, (compare Section A.1), while the latter can be derived from the tapering of the window in time. Obviously δt_{eff} is always smaller than t_{seg} for non-rectangular windows.
As δt_{eff} and δf_{eff} are both defined by the Fourier window, a connection of both values can be found to define suitable x_{OV} values for given window functions. A straightforward criterion can for example be provided by the the main lobe half-width θ or alternatively, the location of the first zero in $\mathcal{F}\{w\}$, at the frequency $\pm f_{\mathrm{z}}$. Here, $t_{z,1} = 1/|f_{\mathrm{z}}|$ can be interpreted as an "effective length" of the window in the time domain. For rectangular windows, $f_{\mathrm{z,R}} = 1/t_{\mathrm{seg}}$ leads to zero minimum overlap, while for the Blackman-Harris window, $f_{\mathrm{z,BH}} \approx 4/t_{\mathrm{seg}}$ calls for overlap factors of $x_{\mathrm{OV}} \approx 0.75$.
Numerically, in cases $x_{\mathrm{OV}} > 0$ the STFT pixels in the t direction become correlated, which is not obvious to the interpreter of the spectrogram and can be misleading. On the other hand, the "clean" choice $x_{\mathrm{OV}} = 0$ would preserve statistical independence in the t-direction, but would also lead to insufficient usage of the underlying time data because of the window function taper.
The proper selection for x_{OV} thus remains based on the intended result, and is typically between 0 and the f_{z}-based values as derived above.

A.3.4 Interpretation of spectrograms

Effectively the spectrogram is a three-dimensional data representation: The power as an intensity is displayed over a frequency and a time axis. This creates the notion of orthogonality between f and t, which of course is not the case because of the reciprocal connection

between the two quantities. As discussed in the previous section, a further correlation along the time axis is introduced by nonzero overlap factors.
This effect gains importance when analyzing signals which exhibit significant variation during t_{seg}. As an example, an instantaneous jump at t_0 in the frequency of a sinusoidal signal from f_1 to f_2 is considered. With the Heaviside function

$$H(t) = \begin{cases} 0 & \text{if } t < 0 \\ 1 & \text{if } t \geq 0 \end{cases} \quad \circ\!\!-\!\!\bullet \quad \frac{1}{2}\delta(f) - \frac{\mathrm{j}}{2\pi f} \quad , \tag{A.19}$$

this can generally be formulated as

$$s(t) = s_1(t)H(t_0 - t) + s_2(t)H(t - t_0) \tag{A.20}$$

with $s_i(t) = \cos(2\pi f_i t)$ and $f_1 \neq f_2$.
An example spectrogram of this mathematical case without additional noise modeling and with realistic numerical values is provided in Figure A.2.
Obviously all sub-spectra of the spectrogram which have no part in $t_0 = 25\,\mu\mathrm{s}$ will show the expected $\delta(f - f_i) * \mathcal{F}\{w\}$ characteristics of a single-side power spectrum, with additive leakage if this is the case. f_1 was deliberately chosen to exactly match a discrete frequency bin by $f_\mathrm{r}/f_1 \in \mathbb{N}$ for perfect leakage-free representation, while $f_\mathrm{r}/f_2 \notin \mathbb{N}$ illustrates the leakage effect. As a consequence, the spectrogram background is mathematically clean for $t < 25\,\mu\mathrm{s}$, but for $t > 25\,\mu\mathrm{s}$ the inhomogeneity introduced by the leakage is obvious. Depending on the phase between window and signal, the leaked "shoulders" in the spectra shift back and forth between $f > f_2$ and $f < f_2$. The spectrum excerpt, however, shows how well the Fourier window handles this effect, which is confined below a level of $-100\,\mathrm{dB}$.
The jump itself yields a more drastic distortion, as is visible in the middle spectrum. Both frequencies naturally appear, but the energy is spread in a broad-band signature over the entire spectrum. This is a direct consequence of the term proportional to $1/f$ in $\mathcal{F}\{H\}$ (A.19), and a clear and general spectral indicator for discontinuous phenomena in the underlying time signal.

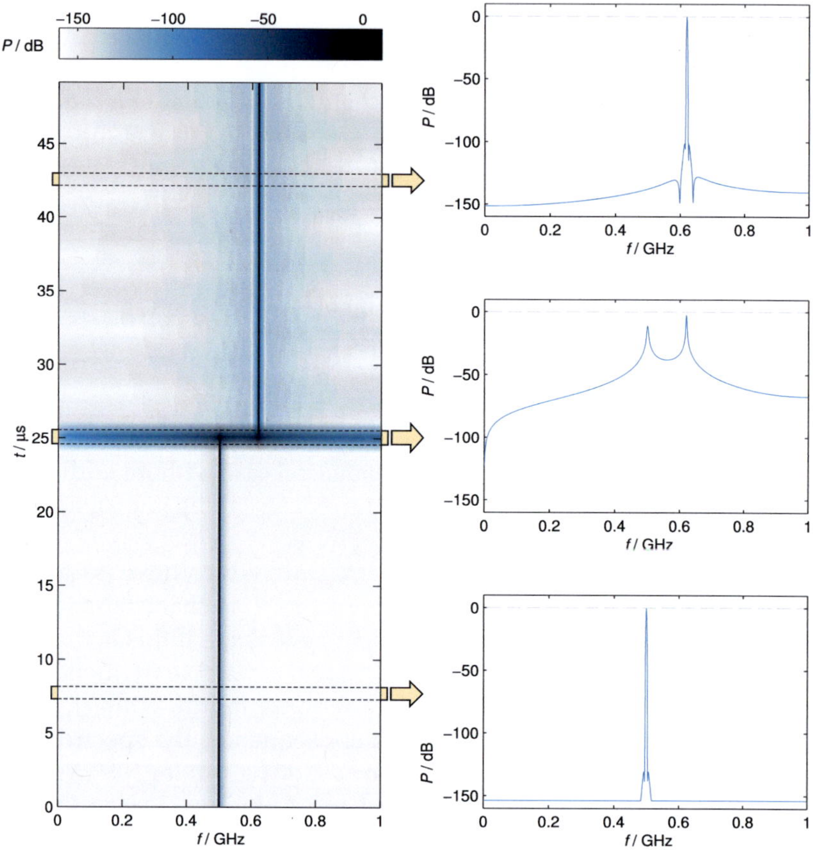

Figure A.2: Numerical sample of a short-time Fourier spectrogram with intermediate spectra. A sinusoidal time signal with $f_1 = 0.5\,\text{GHz}$ in $t \leq 25\,\mu\text{s}$ and $f_2 = 0.62\,\text{GHz}$ in $t > 25\,\mu\text{s}$ with a duration $t_\text{a} = 50\,\mu\text{s}$ is sampled with $f_\text{r} = 2\ \text{GS/s}$. The $1\,\mu\text{s}$-segment STFT with $x_\text{OV} = 0$ and a Blackman-Harris window creates 50 individual spectra.
Left: full spectrogram
Right: single spectra at marked time positions

B PSA system technical supplement

B.1 Control software

The control program was implemented with the integrated development environment (IDE) LABWINDOWS/CVI by National Instruments. This IDE is focused on instrument remote control in general, and facilitates ANSI C based graphical user interface development and maintainance.

Although the PSA functionality is the core application of the control software, it also works as a modular and flexible data acquisition system. For this, a task-based operation scheme was employed: The user defines a list of tasks, which e.g. can be a simple data transfer from an oscilloscope, data retrieval from a modulation domain analyzer (compare Section 1.3.2), or managing a PSA data acquisition, including LO frequency/harmonic management, channel data retrieval, spectrogram generation and RF reconstruction.

A defined list of tasks is processed as soon as the user starts the acquisition cycle, or a defined trigger event is detected. The multi-threaded program processes each task as its own sub-thread; hence, as long as no heavy-data loads occur simultaneously on the same bus from different tasks, the entire acquisition takes approximately as long as the slowest task. For resource-intensive tasks such as the spectrogram generation, the actual task thread can spawn further threads to make optimum use of modern multi-core computers.

The following task types are implemented:

- **PSA**
  Management of one PSA channel pair, i.e. managing the two required LOs, two oscilloscope channels and the ensuing spectrogram generation and RF reconstruction.

- **Filterbank**
  Data download of 8 filterbank channels from the two 4-channel oscilloscopes of the regular frequency measurement system (FMS) [Pri04], including association with the FMS front-end LO necessary for channel interpretation.

- **FTA** (modulation domain analyzer)
  Data acquisition from the FMS modulation domain analyzer, and storage of both relevant LO frequencies.

- **Generic oscilloscope**
  Download of the desired traces from an oscilloscope, without any further functionality.

In Figure B.1 the structure of the named task types is illustrated in a sample setup with a single instance of each task type. Multiple independent instances of each task type are of course possible.

B.2 Data format

Each task also automatically saves the acquired data to the designated measurement file, so when the task processing is finished all data have been stored to the disk. As the output file format, the open TDMS (*technical data management streaming*) standard was chosen, which is also endorsed by National Instruments.
The advantages for the PSA are the binary format with optimization for high read/write performance even for very large file sizes, and having a strict inner structure suited for instrument-centered data organization: Apart from meta-parameters concerning the whole file, it only allows the two hierarchy levels of *Channelgroups* and *Channels*. TDMS files created by the PSA software contain one Channelgroup per task, each Channelgroup containing the complete relevant data associated with the task. For each task, the unique device identifier of the involved instruments and a timestamp for the time of data retrieval is also stored.

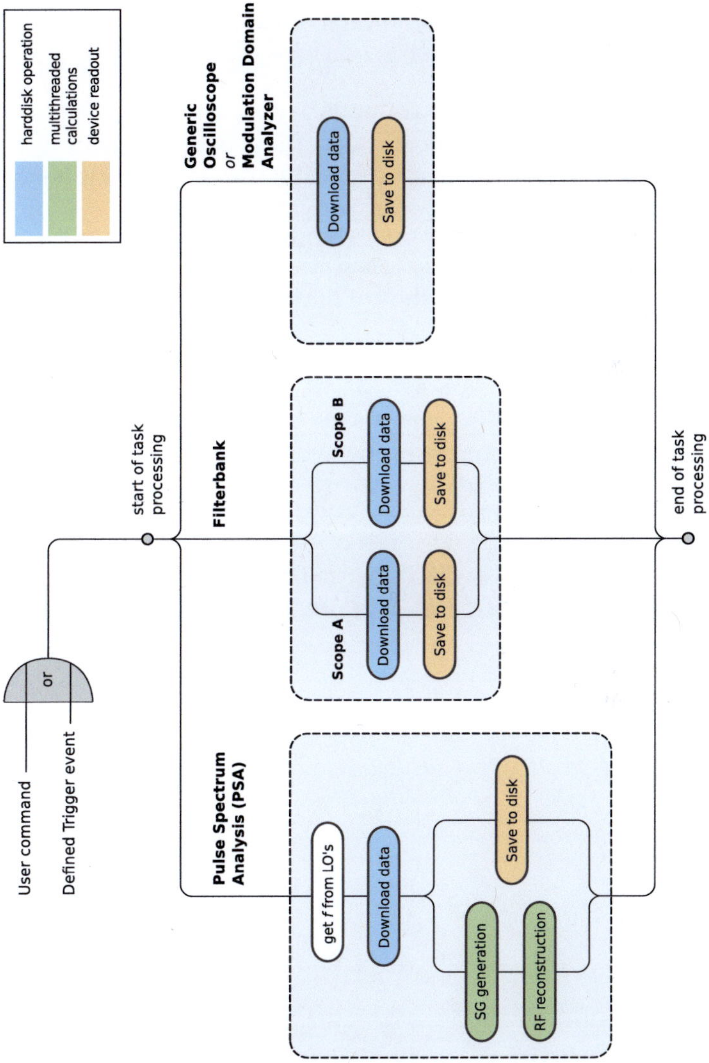

Figure B.1: Parallel task processing example with three separate tasks; generally an arbitrary number of tasks can be defined. Different activities are color-coded.

The following practical possibilities for external processing of these files are available:

- DIADEM, National Instruments' native tool for TDMS analysis; it allows TDMS modification, plotting, batch processing, mathematical analysis and even automatic summary report generation.
- A free plug-in for Microsoft *Excel* [Nat13]; this, however, is only applicable for relatively small files containing channels of less than 32000 entries and thus of limited use for multi-MS size data.
- An add-on named TDMS READER [Hok11] allows the import of TDMS to for Mathworks MATLAB, facilitating advanced scientific analysis. All auxiliary scripts for further processing which were developed in the frame of this work make use of this add-on.

Of course the PSA software is also able to read the TDMS files created, which is particularly useful for fast overviews. The software has extensive plotting capabilities which, particularly for the highly specialized PSA tasks, cannot be performed properly by an external software.

B.3 Components and detailed schematic

In Figure B.2, a more detailed schematic of the PSA system than in Chapter 4 is shown.
It represents the status of the PSA system as established in late 2012, with a total of 4 receiver channels.[1] For clarity, the communication buses and control computer are not included; refer to the *Pulse Spectrum Analysis Control Software Manual* for more a detailed description of these aspects.
Figure B.2 also contains a schematic of the ALC subsystem mentioned in Section 3.1.2. However, the also existing standard frequency measurement system [Pri04] is not drawn.
More information about the individual components can be found in Table B.1, where the manifacturers and the names of the most essential hardware components are listed.

[1]Compare also Figure 4.1, which contains an "elementary cell" of the PSA system.

Function	Manufacturer	Name
Var. attenuator (man.)	Millitech	LSA-06
Var. attenuator (elec.)	Elva-1	VCVA-06
Broadband RF detector	Elva-1	ZBD-06
Waveguide couplers	Elva-1	DC-06/3
Waveguide isolators	Radiometer Physics	WFI WR6.5
Front-end mixers	Virginia Diodes	WR6.5R6 EHM
Local oscillators	PhaseMatrix Agilent	QS FSW20 N5183A MXG
High-pass filters	BSC	EW436
Receiver oscilloscope	Rohde & Schwarz	RTO1024

Table B.1: List of components used in the PSA system

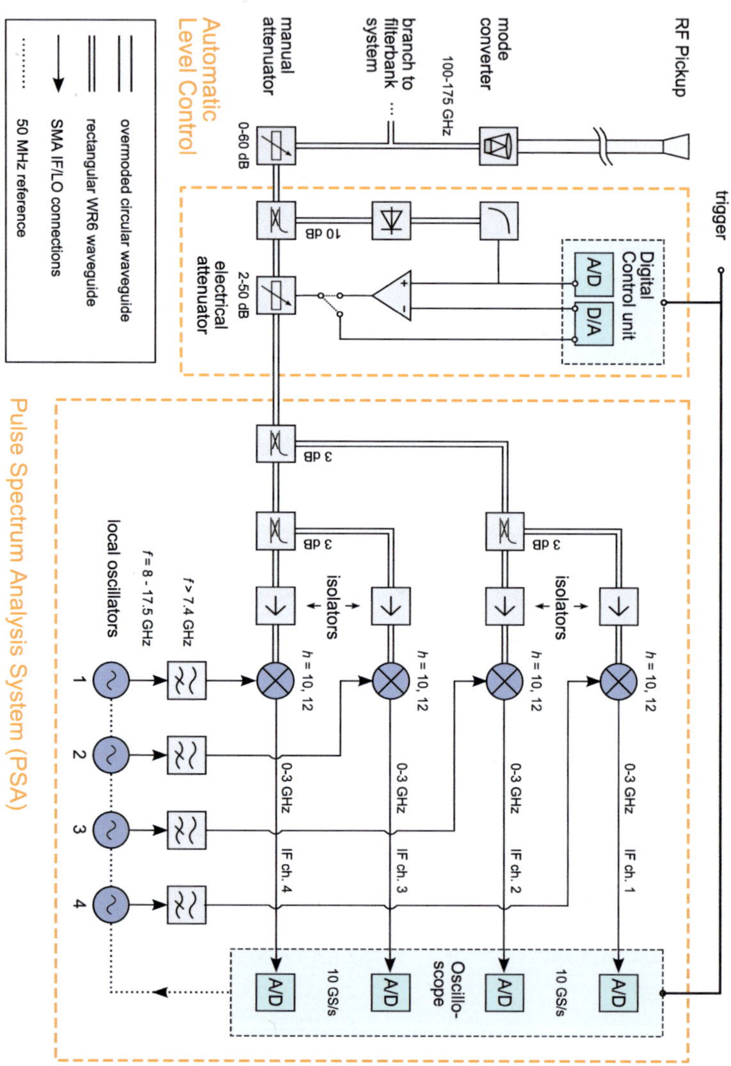

Figure B.2: Overview of spectrum measurement equipment (without filterbank and frequency-time domain analyzer branch)

C Gas kinetics and pressure

This Chapter provides the physical background for the gas calculations presented in Section 7.2. The information was compiled from [Jou08], resp. [JW13], unless stated otherwise. Although all equations are formulated in SI units, for pressure the unit millibar (mbar) instead of Pascal (Pa) is used, to simplify comparisons with the literature.[1]

C.1 Ideal gas conditions and molecular flow rules

C.1.1 Basic relationships

A starting point for analyzing gas behavior is the ideal gas equation

$$p = n\, k_{\mathrm{B}} T$$

$$\begin{aligned} \text{with}\quad & p\text{: pressure in Pascal} \\ & n\text{: particle density in } 1/\mathrm{m}^3 \\ & k_{\mathrm{B}}\text{: Boltzmann constant, } 1.3806488 \cdot 10^{-23}\,\mathrm{J/K} \\ & T\text{: physical temperature in K} \quad . \end{aligned} \tag{C.1}$$

Based on the temperature, the Maxwell-Boltzmann velocity distribution describes the statistical particle speed. For gas particles with the mass m_{p}, the average Brownian motion velocity

$$\bar{v} = \sqrt{\frac{3k_{\mathrm{B}}T}{m_{\mathrm{p}}}} \tag{C.2}$$

follows from the Maxwell-Boltzmann distribution.[2]

[1] $1\,\mathrm{bar} = 10^5\,\mathrm{Pa}$

[2] There also exist other definitions, such as the most probable velocity.

As an example, for nitrogen molecules N_2 with a particle mass of $m_\mathrm{p} = 28\,\mathrm{u} = 4.65 \cdot 10^{-26}\,\mathrm{kg}$, the approximate average velocities for $T_1 = 300\,\mathrm{K}$ and $T_2 = 600\,\mathrm{K}$ are $\overline{v}_1 = 517\,\mathrm{m/s}$ and $\overline{v}_2 = 731\,\mathrm{m/s}$.
The ideal gas equation ignores many effects, such as the kinetics of gas particle collisions, electrostatic polarization forces between the particles, gas friction, turbulence and phase transitions. Therefore it is only applicable to very thin gases, for the description of which these effects are not relevant.
A basic parameter to indicate the rules which apply to the gas in a vessel is the mean free path length, i.e. the average traveling distance of a particle between two collisions with other particles,

$$\bar{l} = \frac{k_\mathrm{B} T}{\sqrt{2}\pi d_\mathrm{p}^2 \cdot p} \quad , \tag{C.3}$$

with the particle diameter d_p. Together with a characteristic size of the pressure vessel d_V, the *Knudsen number*

$$Kn = \bar{l}/d_\mathrm{V} \tag{C.4}$$

is defined from C.3, which separates the different gas flow regimes:

- $Kn < 0.01$, viscous flow
- $0.01 < Kn < 0.5$, transitional flow
- $Kn > 0.5$, molecular flow

An approximate particle size of $0.15\,\mathrm{nm}$ (N_2, taken from [Jou08]) yields $\bar{l} = 41\text{–}83\,\mathrm{m}$ for a typical operation pressure of $p = 10^{-8}\,\mathrm{mbar}$, which is well above the the characteristic gyrotron dimension of $d_\mathrm{V} = 1\,\mathrm{m}$. The resulting minimum Knudsen number at $300\,\mathrm{K}$ of $Kn \approx 21$ is deep in the molecular flow regime.
For approaching the transition regime, a tube pressure of $p \gtrsim 10^{-6}\,\mathrm{mbar}$ would be necessary, which would no longer allow proper operation [Dam95]. For this reason, in the following molecular flow properties are assumed.

C.1.2 Molecular flow

In the molecular flow regime, gas behavior is often counter-intuitive, since many every-day mechanisms generally associated with pressure are not present:

- The occurrence of particle-particle collisions is insignificant. In consequence, there is no internal friction in the gas, no turbulence and no possibility for diffusive effects. Each gas particle "is alone" in the vessel and follows a straight trajectory, which is only interrupted by wall collisions.
- Particles are not reflected from the walls elastically, but are thought to be re-emitted with a uniformly distributed probability for the emission angle with respect to the surface, and a velocity distribution only depending on the wall temperature. Therefore in every wall collision, a gas particle adopts the wall temperature and "loses its history".

C.2 Description of molecular gas flows

The gas flows between different compartments of a system can be described in multiple ways, all of which descend from the ideal gas equation C.1. A basic quantity is so called *pV-Flow* or *throughput*

$$q_{\mathrm{pV}} = p \cdot \dot{V} = N k_{\mathrm{B}} T \quad \text{in } \frac{\mathrm{Pa} \cdot \mathrm{m}^3}{s} \tag{C.5}$$

with the total number of particles N.
From this flow, e.g.

- $q_{\mathrm{V}} = q_{\mathrm{pV}}/p$, volume flow in $\mathrm{m}^3/\mathrm{s} = 10^{-3} \cdot \mathrm{l/s}$
- $q_{\mathrm{N}} = q_{\mathrm{pV}}/(k_{\mathrm{B}} T)$, particle flow in $1/\mathrm{s}$

are derived and converted into each other for the use as fit to the described situation.

C.3 Connections between gas vessels

Typical vacuum problems involve connections between gas vessels, which influence the gas flow from one compartment to the other. It is common to characterize this influence by a conductance value, in a certain analogy to the same-name concept in electrical engineering. This conductance is defined in the form of a volume flow

$$C = \frac{\bar{v}}{4} A P, \quad \text{in } \frac{\mathrm{m}^3}{\mathrm{s}} \tag{C.6}$$

with the temperature-dependent mean velocity $\bar{v}$ (C.2), the geometrical connection aperture A in m^2 and the transmission probability P, which represents the internal geometry of the element. Together with the pressure difference between two vessels, the resulting stationary pV flow

$$q_{\mathrm{pV}} = C\,(p_2 - p_1) \tag{C.7}$$

can be found.

The transmission probability P mainly accommodates the fact that individual particles are re-emitted from the wall in arbitrary directions (compare the statements on molecular flow in C.1), and therefore follow a complicated trajectory until they leave the connection at one end or the other. Thin apertures, i.e. $l \approx 0\,\mathrm{mm}$, are only characterized by their area, hence $P \to 1$.

Cylindrical connections are most common, which can be entirely characterized by the length l and the diameter d. Analytical P expressions exist only for very short ($l \ll d$) and very long ($l \gg d$) connections. Transmission probabilities for circular connections of arbitrary l/d shape can be described by an empirical formula based on Monte-Carlo simulations,

$$P = \frac{14 + 4\,l/d}{14 + 18\,l/d + 3(l/d)^2} \quad , \tag{C.8}$$

which converges to the mentioned analytical expressions for the limit cases $l/d \to 0$ and $l/d \to \infty$.

An important assumption in these probabilities is, that the connected vessels are large, i.e. $\bar{l} \gg d$, which ensures a uniform particle velocity vector distribution at the interfaces. This becomes relevant when series connections of different sizes are analyzed: Simple reciprocal connection of the conductances as done in electrical engineering

$$\frac{1}{C_{\text{tot}}} = \frac{1}{C_1} + \frac{1}{C_2} + \ldots + \frac{1}{C_n} \tag{C.9}$$

can be done; however, this yields only an approximate result: The $n-1$ implicit large vessels between all the elements lower the conductivity result because at each interface, re-establishment of the velocity vector uniformity is assumed.
A better result, which can only be surpassed by particle simulations, is achieved with *Oatley's approach*. Here the aperture ratio of the interfacing components, and the resulting "funneling" effect for the particle velocity vector is taken into account.

C.4 Gas particles and vessel walls

In all vacuum systems, an increase of the structure material temperature causes the gas particle density to rise. This effect is caused by the release of gas from the vessel wall material, and is called *outgassing*.
The process is described by a statistical function which is governed by the temperature. Higher temperatures cause increased release, while for lower temperatures the wall can also act as a particle sink by catching particles.
For the release of particles, the characteristic parameter is the desorption energy W_{d}, defined by the type of bond of the particle to the wall: If the desorption energy is above the thermal energy $k_{\text{B}}T$, almost no net desorption takes place. For $W_{\text{d}} \lesssim k_{\text{B}}T$, an outgassing onset takes place, while for $W_{\text{d}} \ll k_{\text{B}}T$ the particle flow rates easily rise by orders of magnitude (see the instructive comparisons in [Jou08]).[3]

[3]This is the reason for outbaking, without which pressures below 10^{-6} mbar would not be achievable due to weak continuous outgassing of the structure material.

The basic sources of gas particles are

- **Adsorption**
  This describes particles bound to the wall surface, which can happen in two ways:
  - **Physical adsorption** (physisorption)
    The particles are attached to the wall by van-der-Waals forces, forming a thin surface layer with quite low desorption energies $W_\mathrm{d} \lesssim 100\,\mathrm{meV}$.
  - **Chemical adsorption** (chemisorption)
    The binding is of chemical nature and is dissolved by an endothermic reaction. This naturally covers an extremely wide range of energy depending on the vessel material and the involved gas species, which is generally above the energies encountered in physisorption.
- **Absorption**
  Here, particles are stored within the lattice of the bulk material, without having a chemical bond with it. Before the gas arrives at the surface (turning into an adsorbed layer), it is subject to a random walk path inside the bulk, which of course is enhanced by temperature. Especially small particles are candidates for absorption, but also simple molecules are found in relevant quantities in metals.

When components are shelved under atmospheric conditions, the ad- and absorption rates follow the partial pressure of the surroundings.

The process of permeation, which can be viewed as a multi-step phenomenon, also exists: Gas particles from outside a vessel are adsorbed to the outer surface, are then absorbed into the material and reach the inner surface, to be desorbed. For metals, this process is limited to thin walls, very small particles ($\mathrm{H_2}$, He) and elevated temperatures.

C.5 Vacuum pump basics

C.5.1 Pump types

All low-pressure systems include vacuum pumps, which can have very different implementations depending on the intended pressure range, particle species and and throughput. Generally two general families can be separated, which are gas transfer pumps and gas entrapment pumps. The former are typically, but not exclusively, found in the viscous flow regime where they transport large q_{pV} flows from a vessel to an exhaust. All mechanical pumps such as piston pumps, turbo-molecular pumps and also liquid-based pumps belong to the first family.
The latter do not remove the gas from the system formed by vessel and pump, but act as a particle sink by binding the particles. This is done either by adsorbing materials (called "getter materials") in *sorption pumps* or by the use of extremely cold surfaces, which condense all incoming matter, in so-called *cryo-pumps*. These pumps are the key to the high and ultra-high vacuum regime $p < 10^{-5}\,\mathrm{Pa} = 10^{-7}\,\mathrm{mbar}$, but cannot adsorb arbitrarily large numbers of particles.

C.5.2 Characterization

The basic parameter for characterizing a vacuum pump is the *pumping speed* S, which is defined as a volume flow. It is typically constant versus pressure over the operating range. In practice, the pump *throughput* is of interest, which is directly calculated as $q_{\mathrm{pV}} = p \cdot S$. Note that for $S = \mathrm{const}$ a linear dependence on pressure is found.
If a vacuum pump is connected to the vessel with a connecting element of a finite conductivity C, the pumping speed S with pressure p_1 at the pump intake will be reduced to an *effective pumping speed* S_{eff} with pressure p_2 at the vessel outlet. Using equation C.7 in C.3 with the throughput continuity $q_{\mathrm{pV},1} = q_{\mathrm{pV},2}$, this value can be found to be

$$S_{\mathrm{eff}} = \frac{p_1}{p_2} S = \frac{S}{1 + S/C} \quad . \tag{C.10}$$

C.6 Ion getter pumps (IGP)

C.6.1 General information

IGPs use a continuous low-current high-voltage vacuum discharge in the lower kV range, which creates a local plasma from the gas particles in the IGP chamber through electron impact ionization. The positive ions created this way are accelerated toward the IGP anode, where multiple processes cause an entrapment of the particles:

1. Most species are directly chemisorbed by the hot anode surface, which is made of a getter material, such as Titanium. This applies for the main constituents of air, N_2, O_2 and water vapor.

2. Noble gases do not form chemical bonds. The main entrapment effect for these atoms is direct ion implantation into the anode bulk by the accelerating voltage. It is aided by the constant sputtering of anode surface material, which forms layers above.

With pumping efficiency normalized to N_2, common species like oxygen, carbon dioxide, water vapor and hydrocarbons have values in the range 0.6–1. An exception is hydrogen, which due to its low mass is pumped more efficiently with 1.5–2. The problematic noble gases are in the range between 0.1–0.3 for triode IGPs.
It should be noted again that because of the molecular flow behavior, there is no "suction" in the every-day sense from the pump. Its inlet rather acts as a trap for the gas particles, which do not drag each other.

C.6.2 Pressure measurement

Ion getter pumps can be used in parallel as a particle density gauge, which is a highly valuable feature. The IGP power supply acts as a high voltage source to the IGP itself, and can measure the current of the continuous pumping discharge. Externally, the result is translated into a monitor voltage which can be converted to a pressure value by the pump calibration data. We refer to this value as the "displayed pressure".

Although even the manufacturer manuals give a pressure value for the convenience of the user, the physical quantity measured by the pump is always the particle density [Jou08]. If the pump temperature T_{pump} itself deviates from the calibration temperature $T_{\mathrm{norm}} = 300\,\mathrm{K}$, the actual pressure inside the pump chamber p_{pump} differs from the displayed pressure p_{disp} according to

$$p_{\mathrm{pump}} = \frac{T_{\mathrm{pump}}}{T_{\mathrm{norm}}} p_{\mathrm{disp}} \quad . \tag{C.11}$$

Depending on the anode material, the displayed pressure can also vary slightly even between pumps of the same type, however, typically only within a range of less than 10% [Jou08].

C.6.3 W7-X gyrotron pumps

The TH1507 W7-X gyrotrons are each equipped with four ion getter pumps at the mirrorbox, which are in continuous operation to maintain the low pressures required to operate. The pumping speed of the individual pumps S_{IGP} is specified to be 14–16 l/s for operation below 10^{-5} mbar [Var94].
For pressure measurement, the IGP current is converted from the display voltage over

$$I_{\mathrm{pump}} = 10^{-1+\frac{6}{5}V_{\mathrm{disp}}} \tag{C.12}$$

which yields the pressure in mbar:

$$p_{\mathrm{disp}} = 1.33322\, I_{\mathrm{pump}} \cdot \frac{3}{8} \cdot 10^{-8} \tag{C.13}$$

A monitor voltage range 0–5 V thus yields a measurement range of $5 \cdot 10^{-10}$ mbar to $5 \cdot 10^{-4}$ mbar directly at the pump. The pumps are connected to the mirrorbox vessel by means of cylindrical tubes, which contain copper screen shields with optimized thickness and hole diameter for protection against the internal microwave stray radiation. These connectors can be modeled as a tubular connection with a thin partial aperture, yielding a connectivity of $C_{\mathrm{conn}} \approx 110\,\mathrm{l/s}$.

D Extended simulation results

This is a supplement to the benchmark in Chapter 3, where the corresponding mode power evolutions as a function of time are given in the Figures 3.5 and 3.6. The operating point parameters are $V_\mathrm{b} = 77.0\,\mathrm{kV}$, $I_\mathrm{b} = 41.7\,\mathrm{A}$, $\alpha = 1.34$, $\delta\alpha_\sigma = 8\%$ with a peak magnetic flux density of $5.606\,\mathrm{T}$. On the following pages, the associated field patterns and spectra are given as an example, in addition to the results from EVR-PIC.

The field patterns in Figures D.1, D.3 and D.5 show the magnitudes of the transverse electric field envelope as a function of z, combined with a sketch of the azimuthally symmetrical geometry.

Similarly, the spectra in Figures D.2, D.4 and D.6 also show field magnitudes instead of powers. Because the DFT segment lengths are limited by simulation resources and a separate trace is available for each mode, here traditionally rectangular window functions are used to maximize frequency resolution.

It should be noted that not all field/mode normalization factors are included in these types of output, so comparison between tools and even different traces in the same graph is less accurate than in the power vs. time plots.

The three tools agree well in their primary result, but in both types of representation the differences in numerics and implementation of physics is obvious. For example, the noise floors differ in their structure, as in the start-ups presented in Section 3.3.

D.1 SELFT

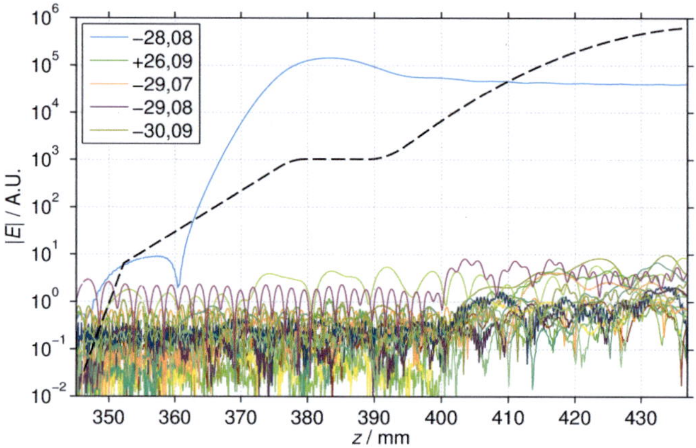

Figure D.1: Fields with SELFT at $V_b = 77.0\,\mathrm{kV}$ (see Section 3.3.4)

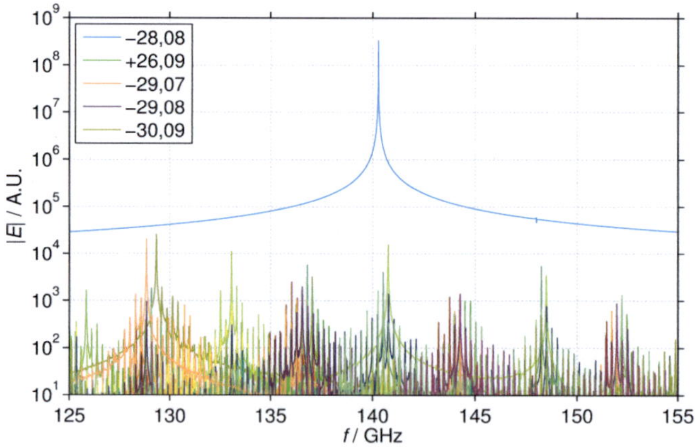

Figure D.2: Amplitude spectra with SELFT at $V_b = 77.0\,\mathrm{kV}$ (see Section 3.3.4)

D.2 EVRIDIKI

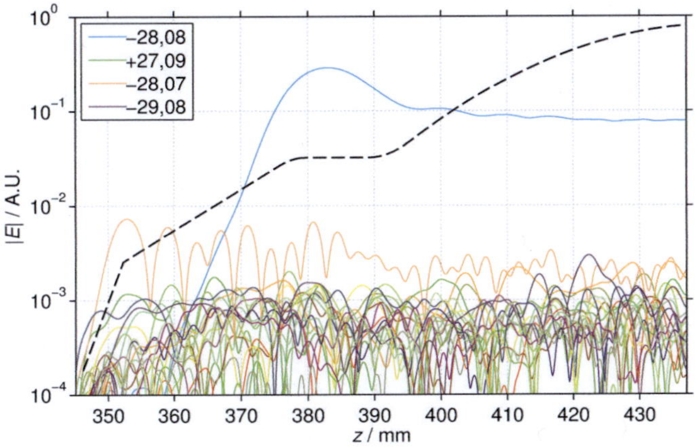

Figure D.3: Fields with EVRIDIKI at $V_\mathrm{b} = 77.0\,\mathrm{kV}$ (see Section 3.3.5)

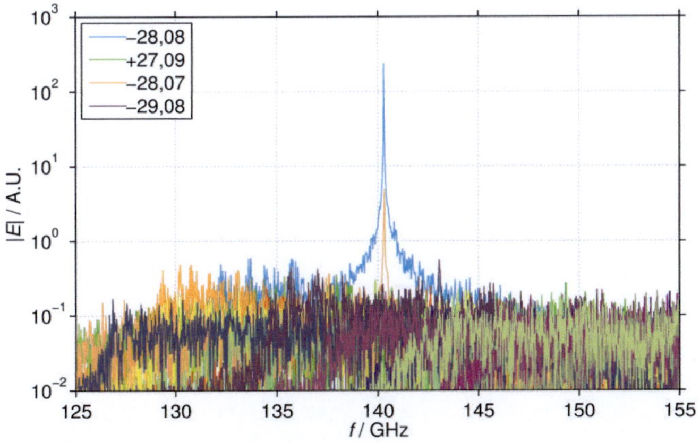

Figure D.4: Amplitude spectra with EVRIDIKI at $V_\mathrm{b} = 77.0\,\mathrm{kV}$ (see Section 3.3.5)

D.3 EVR-PIC

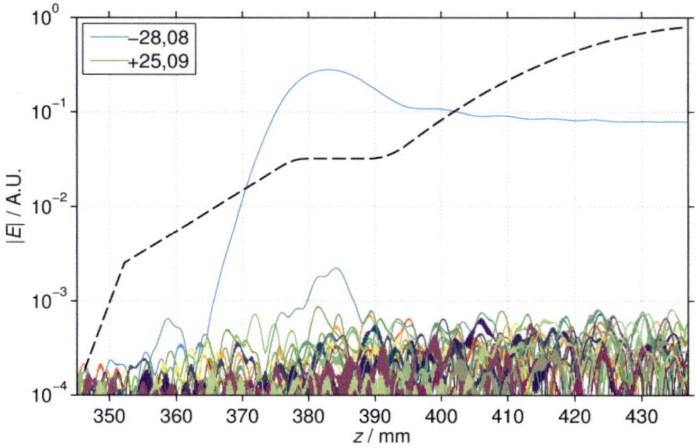

Figure D.5: Fields with EVR-PIC at $V_b = 77.0\,\mathrm{kV}$ (see Section 3.3.5)

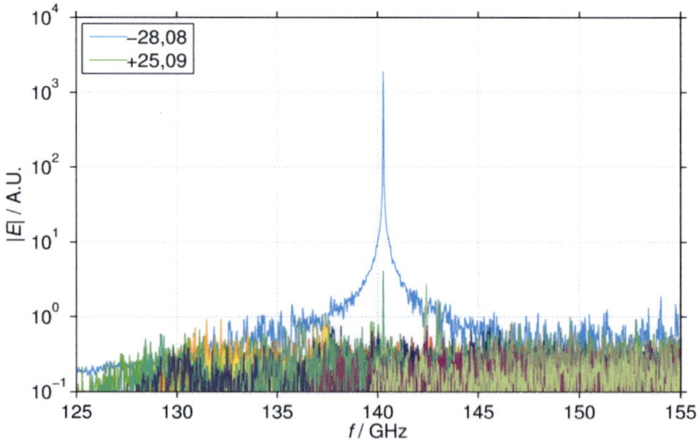

Figure D.6: Spectra with EVR-PIC at $V_b = 77.0\,\mathrm{kV}$ (see Section 3.3.5)

E Afterword

The presented work was accomplished during my time as teaching assistant at the Institut für Hochfrequenztechnik und Elektronik (IHE) and research assistant at the Institute for Pulsed Power and Microwave Technology (IHM) at the Karlsruhe Institute of Technology (KIT).

First of all I want to express my deepest gratitude to Prof. Dr. rer. nat. Dr. h.c. Manfred Thumm for sparking my activity in the field of gyrotrons, the many opportunities he provided me with, and of course for acting as my doctoral advisor. It was the unique combination of freedom and continuous support from his side which made this accomplishment possible.

Further I would like to thank Prof. Dr.-Ing. Dr. h.c. Klaus Schünemann for being my second reviewer, and the kind interest he has shown in my work.

I am deeply grateful to Prof. Dr. Edith Borie for the meticulus and productive proofreading she provided for the thesis.

As I had the great pleasure of also being involved in teaching throughout my whole time at IHE/IHM, I want to express my gratitude to all colleagues involved. Be it the Microwave Measurement Techniques lectures with Prof. Thumm, and later with Dr.-Ing. Mario Pauli, or the new High Power Microwave Technology lectures with Prof. Dr.-Ing. John Jelonnek, I always enjoyed the friendly and stimulating atmosphere of academia.

Here additional thanks go to John Jelonnek for the continuous support, kind guidance and encouragement in the course of my work.

I want to express my deep gratitude to the colleagues at IHM, many of who have contributed to this work in one way or the other, and of course also for the pleasant and productive working environment. There is not room to thank the many who deserve it, but I want them all to know how grateful I am. Special thanks go to the staff of building

691, who never were hesitant in providing a quick fix for the equipment or a nudge in the right direction, and to Andrey Samartsev, who spent innumerable hours operating gyrotrons for my not always successful experiments. Further I am grateful to Stefan Kern, whose kind mentoring during the first years broadened my perspective, and who never failed to suggest inspiring approaches to science and engineering problems, be they theoretical or practical. Jens Flamm was one of the people who got me "hooked up" on gyrotrons in the first place; for this and also the many stimulating discussions on the broadest range of topics during our common time at IHM and beyond and I want to thank him.

The colleagues at IHE always gave me a feeling of affiliation, even though most of my time was spent on KIT Campus North. Especially I want to thank Philipp Pahl, and of course the head of the institute, Prof. Dr.-Ing. Zwick.

Also I would like to thank the secretariats of both institutes, who never failed to show me the right path in the at times thick bureaucratic jungle of teaching, financing and business trips.

Furthermore, I express my appreciation to the various students whose work was performed within the frame of my thesis, starting with Dipl.-Ing. Konstantin Nußbaum for developing the ALC system, which made high power measurements much more efficient and surely saved several millimeter-wave mixers over the years. I am grateful to B.Sc. Helmar Schütz for making numerous contributions, big and small, to different parts of the work over the time, to M.Sc. Chuanren Wu who with great numerical intuition got the fitting technique used in Chapter 7 underway, and to M.Sc. Veenu Kamra who implemented the Brillouin analysis technique described in Section 6.3 with great effort.

I thank Jianghau Zhang for being a very pleasant room mate in my IHM office, and our stimulating discussions.

Outside of KIT, there is also a number of people who I would like to thank explicitly. Of *Thales Electron Devices (TED)*, I hereby want to thank Christophe Liévin, Francois Legrand and Gérald Lietaer for their kind cooperation during FAT and SAT long pulse experiments, and numerous technical discussions. Volker Erckmann, Georg Michel and

the "three Franks" at the *Max Planck Institut für Plasmaphysik (IPP)* in Greifswald, Germany, facilitated the post-SAT measurement campaign on W7-X gyrotron SN4R in 2011 in a great atmosphere. Similar thanks for the given opportunities and the great working environment are directed to the gyrotron staff at the *Centre de Recherches en Physique des Plasmas (CRPP)* of the *École Polytechnique Fédérale de Lausanne* (EPFL), i.e. Stefano Alberti, Jean-Philippe Hogge and Feng-Ping Li.
Very special thanks go to my wife for her patience and support, and to my daughter Johanna Juna for many welcome distractions.

Karlsruhe, May 2015
Andreas Schlaich

Bibliography

[AAA$^+$12] S. Alberti, J.-Ph. Ansermet, K. A. Avramidis, F. Braunmüller, P. Cuanillon, J. Dubray, D. Fasel, J.-Ph. Hogge, A. Macor, E. de Rijk, M. da Silva, M. Q. Tran, T. M. Tran, and Q. Vuillemin. Experimental study from linear to chaotic regimes on a terahertz-frequency gyrotron oscillator. *Physics of Plasmas*, 19(12):123102, 2012.

[AAB$^+$01] S. Alberti, A. Arnold, E. Borie, G. Dammertz, V. Erckmann, P. Garin, E. Giguet, S. Illy, G. Le Cloarec, Y. Le Goff, R. Magne, G. Michel, B. Piosczyk, C. Tran, M.Q. Tran, M. Thumm, and D. Wagner. European high-power CW gyrotron development for ECRH systems. *Fusion Engineering and Design*, 53(1–4):387–397, 2001.

[ADRS01] M. I. Airila, O. Dumbrajs, A. Reinfelds, and U. Strautins. Nonstationary oscillations in gyrotrons. *Physics of Plasmas*, 8(10):4608–4612, 2001.

[AEDJ97] M. E. Austin, R. F. Ellis, J. L. Doane, and R. A. James. Improved operation of the michelson interferometer electron cyclotron emission diagnostic on DIII-D. *Review of Scientific Instruments*, 68(1):480–483, 1997.

[agi08] Agilent Modulation Domain and Time Interval Analyzers - Pulse Analyzer System (PAS) Z2090B-170. Technical overview, Agilent Technologies, 2008.

[agi13a] Agilent Infiniium 90000 Q-series oscilloscopes data sheet 5990-9712EN. Technical report, Agilent Technologies, 2013.

[agi13b] Agilent PSA Series Spectrum Analyzers. Datasheet, Agilent Technologies, 2013.

[AIS+14] K. A. Avramidis, Z. C. Ioannidis, A. Samartsev, A. R. Choudhury, S. Alberti, and I. G. Tigelis. Simulations of dynamic after-cavity interaction in gyrotrons [S.1]. In *9th Int. Workshop on Strong microwaves and Terahertz waves: Sources and applications*, Nizhny Novgorod, Russian Federation, Jul 2014.

[AJCN92] T. M. Antonsen Jr, S. Y. Cai, and G. S. Nusinovich. Effect of window reflection on gyrotron operation. *Physics of Fluids B: Plasma Physics*, 4(12):4131–4139, 1992.

[Alb12] Stefano Alberti. private communication, 2012.

[AMS+11] H. Aripin, S. Mitsudo, I.N. Sudiana, S. Tani, K. Sako, Y. Fujii, T. Saito, T. Idehara, and S. Sabchevski. Rapid sintering of silica xerogel ceramic derived from sago waste ash using sub-millimeter wave heating with a 300 GHz CW gyrotron. *Journal of Infrared, Millimeter, and Terahertz Waves*, 32(6):867–876, 2011.

[APIT13] K. A. Avramidis, I. Gr. Pagonakis, Z. C. Ioannidis, and I. G. Tigelis. Numerical investigations on the effects of electron beam misalignment on beam-wave interaction in a high-power coaxial gyrotron, [Mo P1-34]. In *38th International Conference on Infrared, Millimeter, and Terahertz Waves (IRMMW-THz)*, Mainz, Germany, 2013. IEEE.

[APIV12] K. A. Avramidis, I. Gr. Pagonakis, C. T. Iatrou, and J. L. Vomvoridis. EURIDICE: A code-package for gyrotron interaction simulations and cavity design. *EPJ Web of Conferences*, 32:04016f, 2012.

[APPV07] K. A. Avramidis, I. Gr. Pagonakis, B. Piosczyk, and J. L. Vomvoridis. Possible excitation of radial satellites in high-power gyrotrons, [Wed P5-10]. In *32nd International Con-*

ference on Infrared, Millimeter and Terahertz Waves (IRMMW-THz), number 3510, pages 700–701, Cardiff, UK, 2007.

[ARD+12] K. A. Avramidis, A. K. Ram, O. Dumbrajs, S. Alberti, T. M. Tran, and S. Kern. On the numerical scheme employed in gyrotron interaction simulations. volume 32, page 04017, 2012.

[ATA+11] S. Alberti, T. M. Tran, K. A. Avramidis, F. Li, and J.-Ph. Hogge. Gyrotron parasitic-effects studies using the time-dependent self-consistent monomode code TWANG, [Tu5.15]. In *36th International Conference on Infrared, Millimeter and Terahertz Waves (IRMMW-THz)*, Houston, TX, USA, 2011.

[AWJ+14] K. A. Avramidis, C. Wu, J. Jelonnek, T.-M. Tran, and S. Alberti. Studies on boundary conditions for gyrotron interaction modelling [3P-6]. In *IEEE International Conference on Plasma Science and 20th International Conference on High-Power Particle Beams (ICOPS/BEAMS 2014)*, Washington D.C., USA, May 2014.

[BAL+98] M. Botton, T. M. Antonsen, B. Levush, K. T. Nguyen, and A. N. Vlasov. MAGY: A time-dependent code for simulation of slow and fast microwave sources. *IEEE Transactions on Plasma Science*, 26(3):882–892, 1998.

[Ber11] M. H. Beringer. *Design studies towards a 4 MW 170 GHz coaxial-cavity gyrotron*. Karlsruher Forschungsberichte aus dem Institut für Hochleistungsimpuls- und Mikrowellentechnik, Band 1, PhD thesis, Karlsruhe Institute of Technology (KIT), 2011. ISBN 978-3-86644-663-2.

[BFB+04] M. Blank, K. Felch, P. Borchard, P. Cahalan, S. R. Cauffman, T. S. Chu, and H. Jory. Demonstration of a high-power long-pulse 140 GHz gyrotron oscillator. *IEEE Transactions on Plasma Science*, 32(3):867–876, June 2004.

[BGJ+92] E. Borie, G. Gantenbein, B. Jödicke, G. Dammertz, O. Dumbrajs, T. Geist, M. Hochschild, M. Kuntze, H.-U. Nickel, B. Piosczyk, and M. Thumm. Mode competition using TE03 gyrotron cavities. *International Journal of Electronics*, 72(5-6):687–720, 1992.

[BHK+07] V. S. Bajaj, M. K. Hornstein, K. E. Kreischer, J. R. Sirigiri, P. P. Woskov, M. L. Mak-Jurkauskas, J. Herzfeld, R. J. Temkin, and R. G. Griffin. 250 GHz CW gyrotron oscillator for dynamic nuclear polarization in biological solid state NMR. *Journal of Magnetic Resonance*, 189(2):251 – 279, 2007.

[BIP08] BIPM, IEC, IFCC, ILAC, ISO, IUPAC, IUPAP, OIML. The international vocabulary of metrology—basic and general concepts and associated terms (VIM), JCGM 200: 2008, 2008.

[Bit04] J. A. Bittencourt. *Fundamentals of plasma physics*. Springer, New York, 3. edition, 2004.

[BJ88] E. Borie and B. Jödicke. Comments on the linear theory of the gyrotron. *IEEE Transactions on Plasma Science*, 16(2):116–121, Apr 1988.

[BJ90] E. Borie and B. Jödicke. Rieke diagrams for gyrotrons. *International Journal of Infrared and Millimeter Waves*, 11(2):243–250, 1990.

[BKC99] J. J. Barroso, K. G. Kostov, and R. A. Correa. Electromagnetic simulation of a 32 GHz, TE021 gyrotron. *IEEE Transactions on Plasma Science*, 27(2):384–390, Apr 1999.

[BKK+88] H. J. Barkley, W. Kasparek, H. Kumrič, G. Müller, P. G. Schüller, M. Thumm, R. Wilhelm, and V. Erckmann. Mode purity measurements on gyrotrons for plasma heating at the stellarator W VII-AS. *International Journal of Electronics*, 64(1):21–28, 1988.

[Bor01] E. Borie. Effect of reflection on gyrotron operation. *IEEE Transactions on Microwave Theory and Techniques*, 49(7):1342–1345, 2001.

[BS04] P. L. Bartlett and A. T. Stelbovics. Electron-impact ionization cross sections for elements Z=1 to Z=54. *Atomic Data and Nuclear Data Tables*, 86(2):235 – 265, 2004.

[BSS07] J. Benford, J. A. Swegle, and E. Schamiloglu. *High Power Microwaves*. CRC Press, 2007.

[CA81] J. Y. Choe and S. Ahn. General mode analysis of a gyrotron dispersion equation. *IEEE Transactions on Electron Devices*, 28(1):94–102, 1981.

[CAV13] I. G. Chelis, K. A. Avramidis, and J. L. Vomvoridis. Simulation of parasitic gyrotron interaction in beam tunnels [Mo P1-65]. In *38th International Conference on Infrared, Millimeter, and Terahertz Waves (IRMMW-THz)*, Mainz, Germany, September 2013.

[CBB+11] S. R. Cauffman, M. Blank, P. Borchard, P. Cahalan, and K. Felch. Recent tests on a multi-megawatt 95 GHz gyrotron, [IO6B-2]. In *IEEE International Conference on Plasma Science (ICOPS)*, Chicago, IL, USA, June 2011.

[Che14] I. G. Chelis. private communication, 2014.

[Cho14] A. R. Choudhury. *Investigations of After Cavity Interaction in Gyrotrons Including the Effect of Non-uniform Magnetic Field*. Karlsruher Forschungsberichte aus dem Institut für Hochleistungsimpuls- und Mikrowellentechnik, Band 4, PhD thesis, Karlsruhe Institute of Technology, Karlsruhe, 2014. ISBN 978-3-7315-0129-9.

[Chu04] K. R. Chu. The electron cyclotron maser. *Rev. Mod. Phys.*, 76:489–540, May 2004.

[CMO00] W. W. Camp, J. T. Mayhan, and R. M. O'Donnell. Wideband radar for ballistic missile defense and range-doppler imaging of satellites. *Lincoln Laboratory Journal*, 12(2):267–280, 2000.

[Coh89] L. Cohen. Time-frequency distributions - a review. *Proceedings of the IEEE*, 77(7):941–981, Jul 1989.

[Con83] V. Considine. Digital complex sampling. *Electronics Letters*, 19(16):608–609, 1983.

[Cor13] M. Cordova. Thermo-mechanical study of the cavity of the 1 MW – 140 GHz gyrotron for W7-X. Master's thesis, Karlsruhe Institute of Technology, 2013.

[CSST07] E. M. Choi, M. A. Shapiro, J. R. Sirigiri, and R. J. Temkin. Experimental observation of the effect of aftercavity interaction in a depressed collector gyrotron oscillator. *Physics of Plasmas*, 14(9):–, 2007.

[CV12] I. G. Chelis and J. L. Vomvoridis. Electromagnetic modelling of dielectric loaded aperiodic gyrotron beam tunnels. In *EPJ Web of Conferences*, volume 32, page 04018, 2012.

[DAA+02] G. Dammertz, S. Alberti, A. Arnold, E. Borie, V. Erckmann, G. Gantenbein, E. Giguet, R. Heidinger, J. P Hogge, S. Illy, W. Kasparek, K. Koppenburg, M. Kuntze, H. P. Laqua, G. LeCloarec, Y. LeGoff, W. Leonhardt, C. Lievin, R. Magne, G. Michel, G. Muller, G. Neffe, B. Piosczyk, M. Schmid, K. Schworer, M. Thumm, and M.Q. Tran. Development of a 140 GHz 1 MW continuous wave gyrotron for the W7-X stellarator. *IEEE Transactions on Plasma Science*, 30(3):808–818, June 2002.

[DAA+06] G. Dammertz, S. Alberti, A. Arnold, D. Bariou, P. Brand, H. Braune, V. Erckmann, O. Dumbrajs, G. Gantenbein,

E. Giguet, R. Heidinger, J.-P. Hogge, S. Illy, J. Jin, W. Kasparek, K. Koppenburg, H.P. Laqua, F. Legrand, W. Leonhardt, C. Lievin, G. Michel, G. Neffe, B. Piosczyk, O. Prinz, T. Rzesnicki, M. Schmid, M. Thumm, Minh Quang Tran, X. Yang, and I. Yovchev. High-power gyrotron development at Forschungszentrum Karlsruhe for fusion applications. *IEEE Transactions on Plasma Science*, 34(2):173–186, Apr 2006.

[DAF+03] G. Dammertz, S. Alberti, D. Fasel, E. Giguet, K. Koppenburg, M. Kuntze, F. Legrand, W. Leonhardt, C. Lievin, G. Müller, et al. Power modulation capabilities of the 140 GHz / 1 MW gyrotron for the stellarator Wendelstein 7-X. *Fusion engineering and design*, 66:497–502, 2003.

[Dam95] G. Dammertz. Vacuum requirements in high power microwave tubes. *Vacuum*, 46(8–10):785–788, 1995.

[DBC+99] G. Dammertz, O. Braz, A. K. Chopra, K. Koppenburg, M. Kuntze, B. Piosczyk, and M. Thumm. Recent results of the 1 MW, 140 GHz, TE22,6-mode gyrotron. *IEEE Transactions on Plasma Science*, 27(2):330–339, 1999.

[DBI+96] G. Dammertz, O. Braz, C.T. Iatrou, M. Kuntze, A Mobius, B. Piosczyk, and M. Thumm. Long-pulse operation of a 0.5 MW TE10,4 gyrotron at 140 GHz. *IEEE Transactions on Plasma Science*, 24(3):570–578, Jun 1996.

[DBMM00] H. Deutsch, K. Becker, S. Matt, and T. D. Märk. Theoretical determination of absolute electron-impact ionization cross sections of molecules. *International Journal of Mass Spectrometry*, 197(1):37–69, 2000.

[Dic13] R. M. Dickinson. Power in the sky: Requirements for microwave wireless power beamers for powering high-altitude platforms. *IEEE Microwave Magazine*, 14(2):36–47, 2013.

[DK81] A. T. Drobot and K. Kim. Space charge effects on the equilibrium of guided electron flow with gyromotion. *International Journal of Electronics*, 51(4):351–367, 1981.

[DN13] O. Dumbrajs and G. S. Nusinovich. Effect of electron beam misalignments on the gyrotron efficiency. *Physics of Plasmas (1994-present)*, 20(7), 2013.

[DT03] V. A. Dolgashev and S. G. Tantawi. Effect of RF parameters on breakdown limits in high-vacuum X-band structures. In *AIP Conference Proceedings*, volume 691, Berkley Springs, WV, USA, Sept 2003. SLAC-PUB-10175.

[DZLM03] G.G. Denisov, V.E. Zapevalov, A.G. Litvak, and V.E. Myasnikov. Megawatt gyrotrons for ECR heating and current-drive systems in controlled-fusion facilities. *Radiophysics and Quantum Electronics*, 46(10):757–768, 2003.

[EBB+07] V. Erckmann, P. Brand, H. Braune, G. Dammertz, G. Gantenbein, W. Kasparek, H.P. Laqua, H. Maassberg, N.B. Marushchenko, G. Michel, et al. Electron cyclotron heating for W7-X: Physics and technology. *Fusion Science and Technology*, 52(2):291–312, 2007.

[Edg80] Ch. J. Edgcombe. The dispersion equation for the gyrotron amplifier. *International Journal of Electronics Theoretical and Experimental*, 48(6):471–486, 1980.

[Edg93] Ch. J. Edgcombe, editor. *Gyrotron Oscillators : their principles and practice*. Taylor & Francis, Washington, DC, 1993.

[elv00] Real Time Frequency Analyzer 120–180 GHz - Part No. RTA-D-180. Operation manual, Elva-1, 2000.

[Eng84] M. Engelson. *Modern spectrum analyzer theory and applications*. Artech House, Dedham, Mass., 1984.

[GBD+07] G. Gantenbein, H. Braune, G. Dammertz, S. Alberti, V. Erckmann, J.-P. Hogge, S. Illy, W. Kasparek, H. P. Laqua,

F. Legrand, W. Leonhardt, C. Lievin, G. Michel, G. Neffe, F. Noke, B. Piosczyk, F. Purps, M. Schmid, M. Thumm, and M.Q. Tran. High-power experiments with 140 GHz series gyrotrons for W7-X, [2.2, invited]. In *IEEE International Vacuum Electronics Conference (IVEC 2007)*, Kitakyushu, Japan, May 2007.

[GDF+10] G. Gantenbein, G. Dammertz, J. Flamm, S. Illy, S. Kern, G. Latsas, B. Piosczyk, T. Rzesnicki, A. Samartsev, A. Schlaich, M. Thumm, and I. Tigelis. Experimental investigations and analysis of parasitic RF oscillations in high-power gyrotrons. *IEEE Transactions on Plasma Science*, 38(6):1168–1177, Mar 2010.

[GP10] R. G. Griffin and T. F. Prisner. High field dynamic nuclear polarization - the renaissance. *Physical Chemistry Chemical Physics*, 12(22):5737–5740, 2010.

[GS03] A. Grudiev and K. Schünemann. Nonstationary behavior of a gyrotron in the presence of reflections. *International Journal of Infrared and Millimeter Waves*, 24(4):429–449, 2003.

[GTW91] T. Geist, M. Thumm, and W. Wiesbeck. Contiguous filterbank receiver for a pulsed 140 GHz gyrotron. In M. R. Siegrist, M.Q. Tran, and T.M. Tran, editors, *16th International Conference on Infrared and Millimeter Waves*, volume 1576, pages 274–275, Lausanne, Switzerland, 1991.

[GZ98] M. Yu. Glyavin and V. E. Zapevalov. Reflections influence on the gyrotron oscillation regimes. *International Journal of Infrared and Millimeter Waves*, 19(11):1499–1511, 1998.

[HAB+07] M.A. Henderson, S. Alberti, P. Benin, T. Bonicelli, R. Chavan, D. Campbell, S. Cirant, G. Dammertz, O. Dormicchi, O. Dumbrajs, D. Fasel, T.P. Goodman, R. Heidinger, J.-P. Hogge, W. Kasparek, C. Lievin, B. Piosczyk, E. Poli,

G. Ramponi, G. Saibene, O. Sauter, A. Serikov, G. Taddia, M. Thumm, M.Q. Tran, A.G.A. Verhoeven, and H. Zohm. EU developments of the ITER ECRH system. *Fusion Engineering and Design*, 82(5–14):454–462, 2007.

[Hac06] K. E. Hackett. Directed energy applications for high power vacuum electronics, [PL.5, plenary]. In *IEEE International Vacuum Electronics Conference, held jointly with IEEE International Vacuum Electron Sources (IVEC-IVESC)*, pages 11–13, Monterey, CA, USA, 2006.

[Har78] F. J. Harris. On the use of windows for harmonic analysis with the discrete Fourier transform. *Proceedings of the IEEE*, 66(1):51–83, 1978.

[Har01] R. F. Harrington. *Time-Harmonic Electromagnetic Fields*. IEEE press / Wiley-Interscience, 2001.

[HCM+08] Y. Hidaka, E. M. Choi, I. Mastovsky, M. A. Shapiro, J. R. Sirigiri, and R. J. Temkin. Observation of large arrays of plasma filaments in air breakdown by 1.5 MW, 110 GHz gyrotron pulses. *Physical Review Letters*, 100(3):035003, 2008.

[HCM+09] Y. Hidaka, E. M. Choi, I. Mastovsky, M. A. Shapiro, J. R. Sirigiri, R. J. Temkin, G. F. Edmiston, A. A. Neuber, and Y. Oda. Plasma structures observed in gas breakdown using a 1.5 MW, 110 GHz pulsed gyrotron. *Physics of Plasmas*, 16:055702, 2009.

[Hok11] J. Hokanson. *TDMS Reader*. The Mathworks File Exchange, 2011. (last verified 20.07.2014).

[hp200] Agilent 53310A Modulation Domain Analyzer. Datasheet, Hewlett Packard / Agilent Technologies, 2000.

[IB99] S. Illy and E. Borie. Investigation of beam instabilities in gyrotron oscillators using kinetic theory and particle-in-cell simulation. *Journal of Plasma Physics*, 62(01):95–115, 1999.

[Ill11] S. Illy. private communication, 2011.

[JAF+13] J. Jelonnek, K. A. Avramidis, J. Franck, G. Gantenbein, K. Hesch, S. Illy, J. Jin, A. Malygin, I. Gr. Pagonakis, T. Rzesnicki, A. Samartsev, T. Scherer, A. Schlaich, M. Schmid, D. Strauss, J. Zhang, and M. Thumm. KIT gyrotron development for future fusion applications, [Mo10-3]. In *38th International Conference on Infrared, Millimeter, and Terahertz Waves (IRMMW-THz)*, Mainz, Germany, Sept 2013.

[Jel00] J. Jelonnek. *Untersuchung des Lastverhaltens von Gyrotrons*. Verein Deutscher Ingenieure: Fortschritt-Berichte VDI / 21 ; 297, PhD thesis, TU Hamburg-Harburg, Düsseldorf, 2000. ISBN 3-18-329721-3. German.

[JGS99] J. Jelonnek, A. Grudiev, and K. Schünemann. Rigorous computation of time-dependent electromagnetic fields in gyrotron cavities excited by internal sources. *IEEE Transactions on Plasma Science*, 27(2):374–383, 1999.

[Jou08] K. Jousten, editor. *Handbook of vacuum technology*. Wiley, Weinheim, 2008.

[JW13] K. Jousten and M. Wutz, editors. *Wutz Handbuch Vakuumtechnik : mit 124 Tabellen und 102 Beispielen*. Springer Vieweg, Wiesbaden, 11., überarb. und erw. aufl. edition, 2013. German.

[KAC+10a] S. Kern, K. A. Avramidis, A. R. Choudhury, E. Borie, G. Gantenbein, S. Illy, A. Samartsev, A. Schlaich, and M. Thumm. Different types of after cavity interaction in gyrotrons. In *US-EU-JPN workshop on RF heating technology*, 2010.

[KAC+10b] S. Kern, K. A. Avramidis, A. R. Choudhury, O. Dumbrajs, G. Gantenbein, S. Illy, A. Samartsev, A. Schlaich, and M. Thumm. Simulation and experimental investigations on dynamic after cavity interaction (ACI), [We-P.14]. In *35th International Conference on Infrared Millimeter and Terahertz Waves (IRMMW-THz)*, Rome, Italy, Sept 2010.

[Kam14] V. Kamra. Study and analysis of parasitic oscillations in gyrotrons. Master's thesis, Indian Institute of Technology Roorkee, 2014.

[KBT04] M. V. Kartikeyan, E. Borie, and M. Thumm. *Gyrotrons : high power microwave and millimeter wave technology*. Advanced texts in physics. Springer, Berlin, 2004.

[KDK+01] K. Koppenburg, G. Dammertz, M. Kuntze, B. Piosczyk, and M. Thumm. Fast frequency-step-tunable high-power gyrotron with hybrid-magnet-system. *IEEE Transactions on Electron Devices*, 48(1):101–107, Jan 2001.

[Ker96] S. Kern. *Numerische Simulation der Gyrotron-Wechselwirkung in koaxialen Resonatoren*. Wissenschaftliche Berichte des Forschungszentrums Karlsruhe FZKA 5837, PhD thesis, Universität Karlsruhe (TH), 1996. German.

[Kes08] W. Kester. Aperture time, aperture jitter, aperture delay time - removing the confusion. *Analog Devices, MT-007 Tutorial*, 2008.

[KM88] W. Kasparek and G. Müller. The wavenumber spectrometer—an alternative to the directional coupler for multimode analysis in oversized waveguides. *International Journal of Electronics*, 64(1):5–20, 1988.

[KSF+09] S. Kern, A. Schlaich, J. Flamm, G. Gantenbein, G. Latsas, T. Rzesnicki, A. Samartsev, M. Thumm, and I. Tigelis.

Investigations on parasitic oscillations in megawatt gyrotrons, [T4C01.0132, keynote]. In *34th International Conference on Infrared, Millimeter, and Terahertz Waves (IRMMW-THz)*, Busan, Korea, Sept 2009.

[KST+01] A. Kasugai, K. Sakamoto, K. Takahashi, K. Kajiwara, H. Shoyama, Y. Ikeda, M. Tsuneoka, T. Fujii, T. Kariya, Y. Mitsunaka, and T. Imai. 1 MW and long pulse operation of gaussian beam output gyrotron with CVD diamond window for fusion devices. *Fusion Engineering and Design*, 53(1–4):399–406, 2001.

[KSW08] U. Kiencke, M. Schwarz, and T. Weickert. *Signalverarbeitung : Zeit-Frequenz-Analyse und Schätzverfahren*. Oldenbourg, Munich, 2008. German.

[LAB+02] M. Lennholm, G. Agarici, A. Barbuti, G. Berger-By, F. Bouquey, J. Clary, C. Darbos, R. Dumont, G. Giruzzi, M. Jung, R. Magne, D. Roux, J-L Segui, and X. Zou. ECRH power injection in TORE SUPRA. In *19th Symposium on Fusion Engineering*, pages 329–332, Atlantic City, NJ, USA, 2002.

[Law57] J. D. Lawson. Some criteria for a power producing thermonuclear reactor. *Proceedings of the Physical Society. Section B*, 70(1):6, 1957.

[LCC+00] J. Lohr, P. Cahalan, R. W. Callis, T. S. Chu, T. C. Luce, C. C. Petty, D. Ponce, R. Prater, D. I. Schuster, and S. E. Tsimring. The 110 GHz gyrotron system on DIII-D: gyrotron tests and physics results. *Proc. Strong Microwaves in Plasmas*, page 46, 2000.

[LFT+99] G. Link, L. Feher, M. Thumm, H.-J. Ritzhaupt-Kleissl, R. Bohme, and A. Weisenburger. Sintering of advanced ceramics using a 30 GHz, 10 kW, CW industrial gyrotron. *IEEE Transactions on Plasma Science*, 27(2):547–554, Apr 1999.

[LIT12] G. P. Latsas, Z. C. Ioannidis, and I. G. Tigelis. Dependence of parasitic modes on geometry and attenuation in gyrotron beam tunnels. *IEEE Transactions on Plasma Science*, 40(6):1538–1544, June 2012.

[LND+08] G. J. Linde, M. T. Ngo, B. G. Danly, W. J. Cheung, and V. Gregers-Hansen. WARLOC: A high-power coherent 94 GHz radar. *IEEE Transactions on Aerospace and Electronic Systems*, 44(3):1102–1117, 2008.

[LST11] A. Litvak, K. Sakamoto, and M Thumm. Innovation on high-power long-pulse gyrotrons. *Plasma Physics and Controlled Fusion*, 53(12):124002, 2011.

[LVAT09] G.P. Latsas, J.L. Vomvoridis, K.A. Avramides, and I.G. Tigelis. Beam-wave interaction in corrugated structures in the small-signal regime. *IEEE Transactions on Plasma Science*, 37(10):2020–2030, Oct 2009.

[Lyo11] R. G. Lyons. *Understanding digital signal processing*. Prentice Hall, Upper Saddle River, NJ [u.a], 3. ed. edition, 2011.

[MAC+03] W. M. Manheimer, T. M. Antonsen, J. P. Calame, B. G. Danly, H. P. Freund, and B. Levush. Electron and ion noise in microwave tubes. *IEEE Transactions on Plasma Science*, 31(1):32–39, 2003.

[Mac11] D. J. C. MacKay. *Sustainable energy - without the hot air*. UIT Cambridge, Cambridge, repr. edition, 2011.

[Mal14] A. Malygin. private communication, 2014.

[MGJ56] E. L. Murphy and R. H. Good Jr. Thermionic emission, field emission, and the transition region. *Physical Review*, 102(6):1464, 1956.

[MIP+13] A. Malygin, S. Illy, I. Gr. Pagonakis, B. Piosczyk, S. Kern, J. Weggen, M. Thumm, J. Jelonnek, K. A. Avramides, R. L.

Ives, D. Marsden, and G. Collins. Design and 3-D simulations of a 10 kW/28 GHz gyrotron with a segmented emitter based on controlled porosity-reservoir cathodes. *IEEE Transactions on Plasma Science*, 41(10):2717–2723, 2013.

[MP12] D. Murakami and K. L. Parkin. An overview of the NASA Ames millimeter-wave thermal launch system, [AIAA 2012-3741]. In *48th AIAA/ASME/SAE/ASEE Joint Propulsion Conference & Exhibit*, Atlanta, Georgia, 2012.

[MTT+90] P. Muggli, M. Q. Tran, T. M. Tran, H. Mathews, G. Agosti, S. Alberti, and A. Perrenoud. Effect of power reflection on the operation of a low-Q 8 GHz gyrotron. *IEEE Transactions on Microwave Theory and Techniques*, 38(9):1345–1351, 1990.

[Nas71] E. Nasser. *Fundamentals of gaseous ionization and plasma electronics*. Wiley-Interscience, New York, 1971.

[Nat13] National Instruments. *TDM Excel Add-in for Microsoft Excel*, 2013. (last verified 20.07.2014).

[NSF+12] J. Neudorfer, A. Stock, J. Flamm, F. Hindenlang, G. Gassner, C. D Munz, R. Schneider, and S. Roller. Numerical investigation of high-order gyrotron mode propagation in launchers at 170 GHz. *IEEE Transactions on Plasma Science*, 40(6):1512–1521, June 2012.

[NTP14] G. S. Nusinovich, M. Thumm, and M. Petelin. The gyrotron at 50: Historical overview. *Journal of Infrared, Millimeter, and Terahertz Waves*, 35(4):325–381, 2014.

[Nus81] G. S. Nusinovich. Mode interaction in gyrotrons. *International Journal of Electronics*, 51(4):457–474, 1981.

[Nus85] G. S. Nusinovich. Analysis of the self-modulation instability of gyrotron radiation. *Radiophysics and Quantum Electronics*, 28(10):926–932, 1985.

[Nus04] G. S. Nusinovich. *Introduction to the physics of gyrotrons*. John Hopkins studies in applied physics. Johns Hopkins Univ. Press, Baltimore [u.a.], 2004.

[Nus11] K. Nussbaum. Einrichtung zum Schutz empfindlicher HF-Komponenten von Frequenzmesssystemen vor Überlast. Master's thesis, Karlsruhe Institute of Technology (KIT), 2011. German.

[Ove53] A. W. Overhauser. Polarization of nuclei in metals. *Phys. Rev.*, 92:411–415, Oct 1953.

[Pag13] I. Gr. Pagonakis. private communication, 2013.

[Pag14] I. Gr. Pagonakis. private communication, 2014.

[PAH+98] M. Pedrozzi, S. Alberti, J. P. Hogge, M. Q. Tran, and T. M. Tran. Electron beam instabilities in gyrotron beam tunnels. *Physics of Plasmas (1994-present)*, 5(6):2421–2430, 1998.

[PAI+14] I. Gr. Pagonakis, K. A. Avramidis, S. Illy, P. Kalaria, G. Gantenbein, and J. Jelonnek. Electron beam simulation in the overall gyrotron geometry [S.11]. In *9th Int. Workshop on Strong microwaves and Terahertz waves: Sources and applications*, Nizhny Novgorod, Russian Federation, Jul 2014.

[Par06] K. Parkin. Microwave heat-exchange thruster and method of operating the same, Feb 2006. US Patent 6,993,898.

[PFC+09] J. I. Paley, F. Felici, S. Coda, T. P. Goodman, and the TCV Team. From profile to sawtooth control: developing feedback control using ECRH/ECCD systems on the TCV tokamak. *Plasma Physics and Controlled Fusion*, 51(12):124041, 2009.

[Pio90a] B. Piosczyk. Compensation of the beam space charge and consequences for the design of a gyrotron, [W8.3]. In R. J. Temkin, editor, *15th International Conference on Infrared and*

Millimeter Waves, volume 1514, pages 499–501, Orlando, FL, USA, Dec. 1990.

[Pio90b] B. Piosczyk. Einfluß der Raumladungskompensation auf den Betrieb eines Gyrotrons. Primärbericht 03 04 02 P16D, Institut für Technische Physik, Kernforschungszentrum Karlsruhe GmbH, 1990. German.

[Pri56] W. L. Pritchard. Long-line effect and pulsed magnetrons. *IRE Transactions on Microwave Theory and Techniques*, 4(2):97–110, 1956.

[Pri04] O. Prinz. Aufbau eines breitbandigen Echtzeit-Frequenzmesssystems für mm-Wellen Hochleistungsgyrotrons. Wissenschaftliche Berichte des Forschungszentrum Karlsruhe FZKA 7020, Forschungszentrum Karlsruhe, Karlsruhe, 2004. German.

[Pro07] J. G. Proakis. *Digital Signal Processing : Principles, Algorithms, and Applications*. Pearson Prentice Hall, NJ/US, 2007.

[PV04] I. Gr. Pagonakis and J. L. Vomvoridis. The self-consistent 3D trajectory electrostatic code ARIADNE for gyrotron beam tunnel simulation. In *29th International Conference on Infrared and Millimeter Waves and 12th International Conference on Terahertz Electronics*, pages 657–658, Karlsruhe, Germany, Sept 2004.

[Rei08] M. Reiser. *Theory and design of charged particle beams*. Wiley-VCH, Weinheim, 2., updated and expanded ed. edition, 2008.

[RGI+13] T. Rzesnicki, G. Gantenbein, S. Illy, J. Jelonnek, J. Jin, I. Gr. Pagonakis, B. Piosczyk, A. Schlaich, and M. Thumm. 2 MW, 170 GHz coaxial-cavity short-pulse gyrotron - investigations on electron beam instabilities and parasitic oscillations, [We5-3]. In *38th International Conference on*

Infrared, Millimeter, and Terahertz Waves (IRMMW-THz), Mainz, Germany, Sept 2013.

[Roh14] Rohde & Schwarz. *R&S RTO Digital Oscilloscope - Scope of the art*, PD 5214.2327.12 edition, Feb 2014.

[RPK+10] T. Rzesnicki, B. Piosczyk, S. Kern, S. Illy, J. Jin, A. Samartsev, A. Schlaich, and M. Thumm. 2.2 MW record power of the 170 GHz european pre-prototype coaxial-cavity gyrotron for ITER. *IEEE Transactions on Plasma Science*, 38(6):1141–1149, 2010.

[rs213] Signal and Spectrum Analyzers from Rohde & Schwarz. Product Brochure 5.00, Rohde & Schwarz GmbH & Co. KG, 2013.

[SCG+11] A. Schlaich, A. R. Choudhury, G. Gantenbein, S. Illy, S. Kern, Ch. Lievin, A. Samartsev, and M. Thumm. Examination of parasitic after-cavity oscillations in the W7-X series gyrotron SN4R [Th2A.4]. In *36th International Conference on Infrared, Millimeter, and Terahertz Waves (IRMMW-THz)*, Houston, TX, USA, Sept. 2011.

[Sch09] A. Schlaich. Aufbau und Anwendung eines Systems zur Spektralanalyse von Gyrotronpulsen im Millimeterwellenbereich. Diploma thesis, Karlsruhe Institute of Technology (KIT), Oct. 2009. KIT Scientific Reports 7541. German.

[Sch11] A. Schaefer. The effective number of bits (ENOB) of my R&S digital oscilloscope - 1ER03-1e. Technical report, Rohde & Schwarz GmbH & Co. KG, 2011.

[Sch14a] T. Scherer. private communication, 2014.

[Sch14b] H. Schütz. Frequency tracing and anti-aliasing in gyrotron STFT spectrograms. Bachelor thesis, Karlsruhe Institute of Technology (KIT), Apr 2014.

[scm73] Glipcop - copper dispersion strenghtened with aluminum oxide. Technical report, SCM metal products, inc., 1973.

[SEG+11] M. Schmid, V. Erckmann, G. Gantenbein, S. Illy, S. Kern, C. Liévin, A. Samartsev, A. Schlaich, T. Rzesnicki, and M Thumm. Technical developments at the KIT gyrotron test facility. *Fusion Engineering and Design*, 86(6–8):518 – 521, 2011. Proceedings of the 26th Symposium of Fusion Technology (SOFT-26).

[SFG+10] A. Schlaich, J. Flamm, G. Gantenbein, G. Latsas, S. Kern, T. Rzesnicki, A. Samartsev, M. Thumm, and I. Tigelis. Investigations on parasitic oscillations in megawatt gyrotrons [2.4], **Best Student Paper Award**. In *International Vacuum Electronics Conference (IVEC) CFP10VAM-ART*. Monterey, CA, USA, May 2010.

[SGJT13a] A. Schlaich, G. Gantenbein, J. Jelonnek, and M. Thumm. Simulations of high power gyrotron operation during window arc [Mo P1-41]. In *38th International Conference on Infrared, Millimeter, and Terahertz Waves (IRMMW-THz)*, Mainz, Germany, Sept. 2013.

[SGJT13b] A. Schlaich, G. Gantenbein, J. Jelonnek, and M. Thumm. Transient millimeter-wave signal analysis with unambiguous RF spectrum reconstruction. *IEEE Transactions on Microwave Theory and Techniques*, 61(12):4660–4666, Dec 2013.

[SGKT12] A. Schlaich, G. Gantenbein, S. Kern, and M. Thumm. Systematic observation of time-dependent phenomena in the RF output spectrum of high power gyrotrons [session technology 1], **invited**. In *Proc. 17th Joint Workshop on Electron Cyclotron Emission and Electron Cyclotron Resonance Heating*, Deurne, the Netherlands, May 2012.

[She94] J. Sheffield. The physics of magnetic fusion reactors. *Rev. Mod. Phys.*, 66:1015–1103, Jul 1994.

[SKI+03] K. Sakamoto, A. Kasugai, Y. Ikeda, K. Hayashi, K. Takahashi, S. Moriyama, M. Seki, T. Kariya, Y. Mitsunaka, T. Fujii, and T. Imai. Development of 170 and 110 GHz gyrotrons for fusion devices. *Nuclear Fusion*, 43(8):729, 2003.

[SKT+07] K. Sakamoto, A. Kasugai, K. Takahashi, R. Minami, N. Kobayashi, and K. Kajiwara. Achievement of robust high-efficiency 1 MW oscillation in the hard-self-excitation region by a 170 GHz continuous-wave gyrotron. *Nature physics*, 3(6):411–414, 2007.

[SN09] O. V. Sinitsyn and G. S. Nusinovich. Analysis of after-cavity interaction in gyrotrons. *Physics of Plasmas*, 16(2):–, 2009.

[SNA12] O. V. Sinitsyn, G. S. Nusinovich, and T. M. Antonsen. Stability of gyrotron operation in very high-order modes. *Physics of Plasmas*, 19(6):–, 2012.

[SNR+12] A. Stock, J. Neudorfer, M. Riedlinger, G. Pirrung, G. Gassner, R. Schneider, S. Roller, and C.-D. Munz. Three-dimensional numerical simulation of a 30 GHz gyrotron resonator with an explicit high-order discontinuous-Galerkin-based parallel particle-in-cell method. *IEEE Transactions on Plasma Science*, 40(7):1860–1870, 2012.

[SSH+01] H. Shoyama, K. Sakamoto, K. Hayashi, A. Kasugai, M. Tsuneoka, K. Takahashi, Y. Ikeda, T. Kariya, Y. Mitsunaka, and T. Imai. High-efficiency oscillation of 170 GHz high-power gyrotron at TE31,8 mode using depressed collector. *Japanese Journal of Applied Physics*, 40(8B):L906, 2001.

[Sti88] H. Stickel. Design of a low average power calorimeter for millimetre wave gyrotrons. *International Journal of Electronics*, 64(1):63–76, 1988.

[TBW$^+$09] D. J. Thoen, W. A. Bongers, E. Westerhof, J. W. Oosterbeek, M.R. De Baar, M. A. Van den Berg, V. Van Beveren, A. Burger, A. P. H. Goede, M. F. Graswinckel, B. A. Hennen, and F. C. Schuller. Development and testing of a fast Fourier transform high dynamic-range spectral diagnostics for millimeter wave characterization. *Review of Scientific Instruments*, 80(10):103504–103504–10, 2009.

[TG10] M. Thumm and G. Gantenbein. Device for preventing parasitic oscillations in electron beam tubes (patent WO 2011/006588 A1), 2010.

[Thu01] M. Thumm. MPACVD-diamond windows for high-power and long-pulse millimeter wave transmission. *Diamond and Related Materials*, 10(9–10):1692 – 1699, 2001.

[Thu03] M. Thumm. Progress in gyrotron development. *Fusion Engineering and Design*, 66–68(0):69 – 90, 2003. 22nd Symposium on Fusion Technology.

[Thu14] M. Thumm. State-of-the-art of high power gyro-devices and free electron masers - update 2013. KIT Scientific Reports 7662, Karlsruhe Institute of Technology (KIT), Karlsruhe, 2014.

[TK87] H. Tawara and T. Kato. Total and partial ionization cross sections of atoms and ions by electron impact. *Atomic Data and Nuclear Data Tables*, 36(2):167–353, 1987.

[TK02] M. Thumm and W. Kasparek. Passive high-power microwave components. *IEEE Transactions on Plasma Science*, 30(3):755–786, June 2002.

[Tsi01] S. E. Tsimring. Gyrotron electron beams: Velocity and energy spread and beam instabilities. *26th International Journal of Infrared and Millimeter Waves*, 22(10):1433–1468, 2001.

[TWK98] M. Thumm, W. Wiesbeck, and S. Kern. *Hochfrequenzmeßtechnik : Verfahren und Meßsysteme*. Teubner, Stuttgart, 2., durchges. aufl. edition, 1998. German.

[Var94] Varian. *Instruction manual for 8 liter per second VacIon pump model 911-5000 and 15 liter per second VacIon pump model 911-5011*. Varian Vacuum Products Division, 1994.

[WEO13] P. Woskov, H. Einstein, and K. Oglesby. Application of fusion gyrotrons to enhanced geothermal systems (EGS). *Bulletin of the American Physical Society*, 58, 2013.

[WGL+08] D. Wagner, G. Grünwald, F. Leuterer, A. Manini, F. Monaco, M. Münich, H. Schütz, J. Stober, H. Zohm, T. Franke, M. Thumm, G. Gantenbein, R. Heidinger, A. Meier, W. Kasparek, C. Lechte, A. Litvak, G. G. Denisov, E. M. Chirkov, A. V. Tai, L. G. Popov, V. O. Nichiporenko, V. E. Myasnikov, E. A. Solyanova, S. A. Malygin, F. Meo, and P. Woskov. Status of the new multi-frequency ECRH system for ASDEX upgrade. *Nuclear Fusion*, 48(5):054006, 2008.

[WSC+07] P. Woskov, S. K. Sundaram, D. Cohn, J. E. Surma, and D. A. Lamar. Directed energy melter, May 10 2007. US Patent App. 12/596,585.

[Wu14] C. Wu. Investigations on improving broadband boundary conditions in gyrotron interaction modelling. Master's thesis, Karlsruhe Institute of Technology (KIT), 2014.

[YBD+03] X. Yang, E. Borie, G. Dammertz, R. Heidinger, K. Koppenburg, B. Piosczyk, D. Wagner, and M. Thumm. A CVD-diamond disk brewster window for a frequency step-tunable 1 MW gyrotron. *International Journal of Infrared and Millimeter Waves*, 24(12):2017–2024, 2003.

[ZGG+01] H. Zohm, G. Gantenbein, A. Gude, S. Günter, F. Leuterer, M. Maraschek, J. Meskat, W. Suttrop, and Q. Yu. Neo-classical tearing modes and their stabilization by electron cyclotron current drive in ASDEX Upgrade. *Physics of Plasmas*, 8:2009–2016, 2001.

[Zha14] J. Zhang. private communication, 2014.

[ZIPJ13] J. Zhang, S. Illy, I. Gr. Pagonakis, and J. Jelonnek. Preliminary study on the effects of emitter surface roughness on gyrotron electron beam quality, [Mo P1-70]. In *38th International Conference on Infrared, Millimeter, and Terahertz Waves (IRMMW-THz)*, Mainz, Germany, Sept 2013.

[ZM04] V. E. Zapevalov and M. A. Moiseev. Influence of after-cavity interaction on gyrotron efficiency. *Radiophysics and Quantum Electronics*, 47(7):520–527, 2004.

[ZSS07] G.I. Zaginaylov, V.I. Shcherbinin, and K. Schünemann. Linear theory of electron cyclotron instability of electromagnetic waves in a magnetoactive plasma waveguide. *Plasma Physics Reports*, 33(8):637–645, 2007.

[ZSST06] G.I. Zaginaylov, V.I. Shcherbinin, K. Schuenemann, and M. Thumm. Influence of background plasma on electromagnetic properties of 'cold' gyrotron cavity. *IEEE Transactions on Plasma Science*, 34(3):512–517, June 2006.

[ZZM00] N.A. Zavolsky, V.E. Zapevalov, and M.A. Moiseev. Numerical study of processes in the cavity of the 170 GHz gyrotron for ITER operating at the TE25.10 mode. *International Journal of Infrared and Millimeter Waves*, 21(9):1381–1395, 2000.

[ZZM06] N.A. Zavolsky, V.E. Zapevalov, and M.A. Moiseev. Numerical simulation of dynamic processes in gyrotrons with low-Q cavities. *Radiophysics and Quantum Electronics*, 49(4):275–287, 2006.

[ZZN79] V. E. Zapevalov, I. G. Zarnitsyna, and G. S. Nusinovich. Excitation of parasitic modes that resonate with the first harmonic of the cyclotron frequency in a gyrotron operating on a mode that resonates with the second harmonic. *Radiophysics and Quantum Electronics*, 22(3):254–258, 1979.

Own publications as sole or principal author

[1] A. Schlaich, C. Wu, I. Gr. Pagonakis, K. Avramidis, S. Illy, G. Gantenbein, J. Jelonnek, and M. Thumm. Separation of thermal expansion and beam charge neutralization effects in high power 140 GHz CW gyrotrons [4B-4]. In *41st IEEE International Conference on Plasma Science and the 20th International Conference on High-Power Particle Beams (ICOPS/BEAMS)*, Washington, D.C., USA, May 2014.

[2] A. Schlaich, G. Gantenbein, J. Jelonnek, and M. Thumm. Transient millimeter-wave signal analysis with unambiguous RF spectrum reconstruction. *IEEE Transactions on Microwave Theory and Techniques*, 61(12):4660–4666, Dec 2013.

[3] A. Schlaich, G. Gantenbein, J. Jelonnek, and M. Thumm. Simulations of high power gyrotron operation during window arc [Mo P1-41]. In *38th International Conference on Infrared, Millimeter, and Terahertz Waves (IRMMW-THz)*, Mainz, Germany, Sept. 2013.

[4] A. Schlaich, G. Gantenbein, J. Jelonnek, A. Samartsev, and M. Thumm. Study of RF output window arc event in a megawatt class 140 GHz gyrotron [10E-8]. In *40th IEEE International Conference on Plasma Science (ICOPS)*, San Francisco, CA, USA, June 2013.

[5] A. Schlaich, J. Jelonnek, and M. Thumm. Millimeter-wave time-domain spectrum analysis system with unambiguous RF spectrum reconstruction [WEPN-1]. In *IEEE International Microwave Symposium*, Seattle, WA, USA, June 2013. IEEE.

[6] A. Schlaich, G. Gantenbein, S. Kern, and M. Thumm. Recent improvements in time-variant gyrotron RF output spectrum monitoring [4b-4], **Best Student Paper Award - Honourable Mention**.

In *39th IEEE Internatioanl Conference on Plasma Science (ICOPS)*, Edinburgh, UK, July 2012.

[7] A. Schlaich, G. Gantenbein, S. Kern, and M. Thumm. Dynamic spectral measurements on high-power oscillators in the millimeterwave domain [Paper No. 1503]. In *Proc. German Microwave Conference 2012*, Ilmenau, Germany, Apr. 2012.

[8] A. Schlaich, G. Gantenbein, S. Kern, and M. Thumm. Systematic observation of time-dependent phenomena in the RF output spectrum of high power gyrotrons [session technology 1], **invited**. In *Proc. 17th Joint Workshop on Electron Cyclotron Emission and Electron Cyclotron Resonance Heating*, Deurne, the Netherlands, May 2012.

[9] A. Schlaich, A. R. Choudhury, G. Gantenbein, S. Illy, S. Kern, Ch. Lievin, A. Samartsev, and M. Thumm. Examination of parasitic after-cavity oscillations in the W7-X series gyrotron SN4R. In *US-EU-JPN RF Heating Technology Workshop*, Austin, TX, USA, Sept. 2011.

[10] A. Schlaich, A. R. Choudhury, G. Gantenbein, S. Illy, S. Kern, Ch. Lievin, A. Samartsev, and M. Thumm. Examination of parasitic after-cavity oscillations in the W7-X series gyrotron SN4R [Th2A.4]. In *36th International Conference on Infrared, Millimeter, and Terahertz Waves (IRMMW-THz)*, Houston, TX, USA, Sept. 2011.

[11] A. Schlaich, J. Flamm, G. Gantenbein, S. Kern, A. Samartsev, and M. Thumm. Characterization of undesired RF oscillations in megawatt gyrotrons. In *International Vacuum Electronics Workshop (IVEW)*, Bad Honnef, Germany, Nov. 2010. L 1.4-2.

[12] A. Schlaich, J. Flamm, G. Gantenbein, G. Latsas, S. Kern, T. Rzesnicki, A. Samartsev, M. Thumm, and I. Tigelis. Investigations on parasitic oscillations in megawatt gyrotrons [2.4], **Best Student Paper Award**. In *International Vacuum Electronics Conference (IVEC) CFP10VAM-ART*. Monterey, CA, USA, May 2010.

[13] A. Schlaich, J. Flamm, G. Gantenbein, S. Kern, G. Latsas, T. Rzesnicki, A. Samartsev, M. Thumm, I. Tigelis, and T. Zwick. Erweiterung der Gyrotron-Frequenzmesstechnik. In *Treffen des Kompetenzbereichs Systeme und Prozesse KIT Karlsruhe*, Karlsruhe, Germany, Mar. 2010. German.

[14] A. Schlaich. Aufbau und Anwendung eines Systems zur Spektralanalyse von Gyrotronpulsen im Millimeterwellenbereich. Diploma thesis, Karlsruhe Institute of Technology (KIT), Oct. 2009. KIT Scientific Reports 7541. German.

[15] A. Schlaich and F. Meyer. A hybrid method for automatic ship detection in SAR imagery. In *Conference on Earth Observing Satellites (CEOS)*. DLR Oberpfaffenhofen, Nov. 2008.

Further journal contributions

[16] G. Gantenbein, A. Samartsev, G. Aiello, G. Dammertz, J. Jelonnek, M. Losert, A. Schlaich, Th. Scherer, D. Strauss, M. Thumm, and D. Wagner. First operation of a step-frequency tunable 1 MW gyrotron with a diamond brewster angle output window. *IEEE Transactions on Electron Devices*, 61(6):1806–1811, June 2014.

[17] M. Schmid, A. R. Choudhury, G. Dammertz, V. Erckmann, G. Gantenbein, S. Illy, J. Jelonnek, S. Kern, F. Legrand, T. Rzesnicki, A. Samartsev, A. Schlaich, and M. Thumm. Recent achievements on tests of series gyrotrons for W7-X and planned extension at the kit gyrotron test facility. *Fusion Engineering and Design*, 88(6):945–949, Oct. 2013.

[18] G. Gantenbein, V. Erckmann, S. Illy, S. Kern, W. Kasparek, C. Lechte, W. Leonhardt, Ch. Liévin, A. Samartsev, A. Schlaich, M. Schmid, and M. Thumm. 140 GHz, 1 MW CW gyrotron development for fusion applications–progress and recent results. *Journal of Infrared, Millimeter, and Terahertz Waves*, 32(3):320–328, Dec. 2011.

[19] M. Schmid, V. Erckmann, G. Gantenbein, S. Illy, S. Kern, Ch. Liévin, A. Samartsev, A. Schlaich, T. Rzesnicki, and M. Thumm. Technical developments at the KIT gyrotron test facility. *Fusion Engineering and Design*, 86(6):518–521, Oct. 2011.

[20] G. Gantenbein, G. Dammertz, J. Flamm, S. Illy, S. Kern, G. Latsas, B. Piosczyk, T. Rzesnicki, A. Samartsev, A. Schlaich, M. Thumm, and I. Tigelis. Experimental investigations and analysis of parasitic RF oscillations in high-power gyrotrons. *IEEE Transactions on Plasma Science*, 38(6):1168–1177, June 2010.

[21] T. Rzesnicki, B. Piosczyk, S. Kern, S. Illy, Jianbo Jin, A. Samartsev, A. Schlaich, and M. Thumm. 2.2 MW record power of the 170 GHz european preprototype coaxial-cavity gyrotron for ITER. *IEEE Transactions on Plasma Science*, 38(6):1141–1149, June 2010.

Index

A

adiabatic approximation 18
AFDR 72
after-cavity osc. (ACI) 117, 121
aliasing 93, 221

B

beam
  radius 16
  voltage 19
beam parameters 17
  simulation 45
  spread 17, 118
beam tunnel 26
before-cavity osc. (BCI) .. 117, 119
Brillouin diagram .... 23 f, 143
bunching 23

C

cathode 15
cathode voltage 19
cavity 20, 26
CAVITY 54
cavity potential well 188
channel pair 77
chemisorption 242
co- / counter-rotation 22
compression zone 26
conductance (gas kinetics) 240
continuation run 55
coupling coefficient 22
  relative 57
cut-off frequency 21
cut-off section 27

D

depression voltage 19
desired acting harmonic 77
desired LO freq. difference 78
DFT processing gain 93, 225
dispersion relation 20, 22
displayed pressure 244
downtaper 26

E

ECRH 3
electron beam .... *see* beam
electron cyclotron frequency 3
electron impact ionization 179
EURIDICE / EVRIDIKI 54

H

heterodyne conv. / mixing 8
high-side injection 9

I

ion getter pumps .... 244 f

K

kinetic energy ... *see* beam voltage

L

Larmor radius ... 17
launcher ... 27
local oscillator ... 8
long-line effect ... 33, 99
long-pulse operation ... 30, 99 ff
low-side injection ... 9

M

major/minor parasites ... 134
mode
    competition ... 28
    propagation ... 20
    starting current ... 29
    switch ... 107, 112
molecular flow ... 239

N

neutralization time ... 194
Nyquist zones ... 94, 221

O

outgassing ... 178, 205, 241
overlap factor ... 227

P

parasitic oscillations ... 7, 117
particle flow ... 239
permeation ... 242
physisorption ... 242
pitch factor ... 16
pulse spectrum analysis ... 13, 69
pumping speed, effective ... 243

R

RF reconstruction ... 77
RMS spread ... 18

S

SELFT ... 54
short-pulse operation ... 30
short-time Fourier transform . 226
slow-time formalism ... 49
space charge ... 19, 185
span ... 11, 75
spectrogram ... 227
spectrum ... 222
spent electron beam ... 25, 121
start-up scenario ... 29, 56

T

throughput ... 239

U

uptaper ... 26

V

voltage depression
    effect ... 19
    neutralization ... 32, 176
volume flow ... 239

W

W7-X ... 5
window function (DFT) ... 222

KARLSRUHER FORSCHUNGSBERICHTE AUS DEM
INSTITUT FÜR HOCHLEISTUNGSIMPULS- UND MIKROWELLENTECHNIK
(ISSN 2192-2764)

Band 1 **MATTHIAS BERINGER**
Design Studies towards a 4 MW 170 GHz Coaxial-Cavity Gyrotron. 2011
ISBN 978-3-86644-663-2

Band 2 **JENS FLAMM**
Diffraction and Scattering in Launchers of
Quasi-Optical Mode Converters for Gyrotrons. 2012
ISBN 978-3-86644-822-3

Band 3 **MATTIA DEL GIACCO**
Investigation of Fretting Wear of Cladding Materials in Liquid Lead. 2013
ISBN 978-3-86644-960-2

Band 4 **AMITAVO ROY CHOUDHURY**
Investigations of After Cavity Interaction in Gyrotrons
Including the Effect of Non-uniform Magnetic Field. 2013
ISBN 978-3-7315-0129-9

Band 5 **MICHAEL BETZ**
The CERN Resonant WISP Search (CROWS). 2014
ISBN 978-3-7315-0199-2

Band 6 **ANDREAS SCHLAICH**
Time-dependent spectrum analysis of high power gyrotrons. 2015
ISBN 978-3-7315-0375-0